T0210747

Competing Discourses on Japan's Nuclear Power

This book examines the discursive formation of nuclear power in Japan to provide insights into the ways this technology has been both promoted and resisted, constituting and being constituted by Japan's sociocultural landscape.

Each chapter pays close attention to a particular discursive site, including newspaper editorials, public relations campaigns, local site fights, urban antinuclear activism, and post-Fukushima pronuclear and antinuclear articulations. The book also raises the question of democracy and sustainability through the examination of nuclear power discourses. It demonstrates the power of discourse in shaping nuclear power by creating knowledge, influencing decisions, relationships, identity, and community. Readers will gain a range of insights from the book: prominent articulations on nuclear power discourse, state and corporate strategies for enticing consent for controversial facilities and technologies, the power of the media in framing public knowledge, the role of social movements and activisms in civic society, the power of community, and nuclear power as a problematic in representative democracy and sustainability.

This book will appeal to students and scholars interested in social discourse, social movements, Japanese society, cultural studies, environmental communication, media analysis, energy and sustainability, and democracy, among others.

Etsuko Kinefuchi is Associate Professor in the Department of Communication Studies and affiliated faculty in the Department of Geography, Environment, and Sustainability, University of North Carolina at Greensboro, USA.

Routledge Studies in Environmental Communication and Media

Journalism, Politics, and the Dakota Access Pipeline
Standing Rock and the Framing of Injustice
Ellen Moore

Environmental Literacy and New Digital Audiences
Patrick Brereton

Reporting Climate Change in the Global North and South
Journalism in Australia and Bangladesh
Jahnnabi Das

Theory and Best Practices in Science Communication Training
Edited by Todd P. Newman

The Anthropocene in Global Media
Neutralizing the Risk
Edited by Leslie Sklair

Communicating Endangered Species
Extinction, News and Public Policy
Edited by Eric Freedman, Sara Shipley Hiles and David B. Sachsman

Communicating Climate Change
Making Environmental Messaging Accessible
Edited by Juita-Elena (Wie) Yusuf and Burton St. John III

Competing Discourses on Japan's Nuclear Power
Pronuclear versus Antinuclear Activism
Etsuko Kinefuchi

For more information about this series, please visit: www.routledge.com/Routledge-Studies-in-Environmental-Communication-and-Media/book-series/RSECM

Competing Discourses on Japan's Nuclear Power

Pronuclear versus Antinuclear Activism

Etsuko Kinefuchi

Routledge
Taylor & Francis Group

LONDON AND NEW YORK

earthscan
from Routledge

First published 2022
by Routledge
2 Park Square, Milton Park, Abingdon, Oxon OX14 4RN

and by Routledge
605 Third Avenue, New York, NY 10158

Routledge is an imprint of the Taylor & Francis Group, an informa business

© 2022 Etsuko Kinefuchi

British Library Cataloguing-in-Publication Data
A catalogue record for this book is available from the British Library

Library of Congress Cataloging-in-Publication Data
A catalog record for this book has been requested

ISBN: 978-0-367-49049-2 (hbk)
ISBN: 978-1-032-15510-4 (pbk)
ISBN: 978-1-003-04422-2 (ebk)

DOI: 10.4324/9781003044222

Typeset in Sabon
by Apex CoVantage, LLC

Contents

Figures

Acknowledgments

This book has my name, but as with any creation, it was not created by me alone. I received support in many beautiful forms.

First, I thank all people who took their time to talk to me during my fieldwork in Japan. I visited multiple communities and sites where antinuclear movements and activisms took place or are still ongoing, including Tokyo, Osaka, Kyoto, Niigata, Aomori, Yamaguchi, and Wakayama. I was not able to feature all of these localities and stories in this book. However, they are all remarkable stories and embodiments of spirited citizenry. I thank them for their courage and perseverance. Among the interviewees, I am especially indebted to Aileen Miyoko Smith of Green Action not only for her time and rich insights but also for introducing me to her network of activists across Japan.

Institutional support was also critical. Fieldwork across Japan was expensive. I am thankful for the Summer Excellence Grant and the Regular Faculty Grant that I received from the University of North Carolina Greensboro (UNCG) for financing the travels that were essential to writing this book. This project also required archival research, which I originally planned to conduct during another trip to Japan. Although the trip was cancelled due to the pandemic, I was able to access the archives I needed thanks to the expert assistance from university libraries. I thank UNCG librarian Jenny Dale for connecting me to Duke University and Duke University librarian Kristina Troost for graciously sponsoring me for guest access to the archives.

I am grateful to Routledge editor, Annabelle Harris, for listening to my ideas, encouraging me to submit a proposal, and reassuringly assisting me through the proposal process. Many thanks go to Routledge editorial assistant, Matthew Shobbrook, for his patience and always timely help with my questions as I labored through the writing process. My gratitude goes also to the anonymous reviewers who provided me with invaluable and generous feedback on my book proposal.

Writing about a topic that is not at the forefront of people's minds can be a lonely process. I am fortunate to have empathetic colleagues and friends who provided me with the moral support that I needed over the years. In particular, I would like to acknowledge my dear friend Lily Mendoza who encouraged me to trust my voice and always gifted me with the perfect words of affirmation.

My daughters, Mina and Maya, were most patient throughout the project. They have grown to take more responsibilities around the house as I was home but often absent from their lives while I was consumed by writing. I truly missed spending time with them. My sisters, Noriko and Yasuko, provided me with much-needed emotional support and kept me connected to my roots in Japan. My heartfelt gratitude goes to my parents who made my education possible and nurtured their disobedient daughter.

Finally, this book would not have been possible without extensive support from my life partner, Todd. I am deeply indebted to him for his assistance with the project. He was my close reader and proofreader. He was also the most patient listener with whom I was able to bounce many ideas. Conversations with him were immensely helpful in refining my writing. Throughout the project, he was my main moral support. Besides all the help he provided with the book, he also brilliantly managed single parenthood while I was away for research for weeks. I am forever grateful.

1 Introduction

It has been ten years since the Fukushima Daiichi nuclear disaster occurred. The tsunami caused by magnitude 9 earthquake on March 11, 2011, in the Tohoku region of Japan disabled the cooling system of the nuclear power plant and caused hydrogen explosions at three of the six reactors at the plant. The explosions at No. 3 reactor were particularly etched in my mind. Split-second orange blasts. Grey round-headed smoke that shot up into the sky. Aghast, I played the videos of the explosions over and over again. It was hard to believe that what I was seeing was really happening. Nuclear power almost never crossed my mind until the accident, and I was disturbed by that realization. I grew up in Niigata, a prefecture adjacent to Fukushima. Niigata is home to the Kashiwazaki-Kariwa Nuclear Power Plant (KNPP), the largest in the world in terms of outputs at the time of the Fukushima accident. The KNPP is less than 30 miles from my home. If a serious accident occurs at the KNPP, part of my home city will have to evacuate. This was always a possibility, but I had never thought about far-reaching consequences of nuclear accidents. I vaguely remember the news about the Chernobyl accident. I'm sure I read and watched something about Japan's nuclear power, too. Yet they did not register in my mind. Why is that?

The Fukushima nuclear disaster became a catalyst for the average Japanese to learn about the pervasiveness of nuclear power in their country. A group of middle-aged women I met at one of the weekly demonstrations in front of the Prime Minister's residence exclaimed, "Did you know that there are 54 nuclear reactors in Japan? We had no idea until this accident happened. And we are such an earthquake-prone country!" Like these women, many citizens were flabbergasted by the sheer number of nuclear power plants spread across Japan. Naively, this was my reaction as well. The pervasiveness of this lack of consciousness is not accidental. It *became* an integral part of Japanese society. It *became* a hegemonic existence. I grew interested in learning about the social and discursive processes that aided the development. As I continued my research into the discourses of nuclear power, however, I met a veteran antinuclear activist, Aileen, who helped me to see the state of this nuclear nation from a different perspective. Aileen has advocated for nuclear-free Japan since the early 1990s. In my interview, I asked

DOI: 10.4324/9781003044222-1

her why we let so many nuclear reactors be built. She replied, "Well, true, we have a lot. But if it wasn't for the communities that fought to keep more reactors from being built, we would have had lots more. We would have been another France." In fact, more than 30 nuclear power plant projects were cancelled between 1960 and 2013 due to antinuclear protests (Behling, Behling, Williams, & Managi, 2019; Yamaaki, 2012). Aileen's words were my "aha" moment. Until then, I was only focused on how nuclear power became hegemonic in Japan. But there were whole other questions I was not asking. What occurred in these communities? Who were fighting, and how were they able to stop the construction of nuclear power plants? What struggles occurred in these communities and urban areas? What did urban activists do? And there it is. My journey down the rabbit hole of Japan's nuclear power began.

The Fukushima disaster and subsequent antinuclear movements became popular topics for academic scrutiny. In addition to numerous scientific and technical analyses of the potential causes of the accident and implications for the future use of nuclear technology, copious sociological and political science analyses have been published to make sense of social and cultural implications of the disaster and the movements (e.g., Arce & Rice, 2019; Behling et al., 2019; Brown, 2018; Chiavacci & Obinger, 2018; Steinhoff, 2018; Tamura, 2018). After ten years from the disaster and after hundreds of publications, do we need more writings about Fukushima? I offer a two-part response. First, this book is not solely about Fukushima. My time-scale interest spans the entire tenure of Japan's nuclear power history. I am interested in how nuclear power has been imagined, socially embedded, and struggled over. Nuclear power did not suddenly become a reality in Japan due to the disaster. It *became* an integral part of the society, and every step of that integration, there was resistance. If we are to know what is going on today, we must know the past. The social history of nuclear power in Japan is the struggle between two incommensurable paradigms of nuclear power. The nuclear-industrial complex had succeeded in establishing nuclear power as an integral and vital aspect of society. However, with little financial capital, people and communities fought against it from the very start and have persevered in many cases. Howard Zinn (2007) wrote that history should disclose "those hidden episodes of the past when, even if in brief flashes, people showed their ability to resist, to join together, and occasionally to win" (p. 12). Knowing the past resistance, as Zinn argued, is critical if history is of value for imagining a possible future. It is important to remember that the massive and widespread antinuclear protests in the wake of the Fukushima nuclear disaster stand on the road paved by the resilient antinuclear movements that existed long before the disaster.

Second, my analyses are guided by a communication perspective. A communication perspective can give insights into social issues by approaching the system of interaction as textual (Deetz, 1992). That is, identity, behavior, relationships, social structure, values, and beliefs are all understood as communicative formations, and they in turn help to shape other formations. Hence, a communication inquiry offers "descriptions of the various forms of interactives, the processes by which the elements produce, are produced and reproduce structural configurations" (Deetz, 1992, p. 82). The Fukushima disaster garnered considerable attention in Communication Studies. Diverse investigations have taken place to date, including but not limited to TEPCO's public communication (Cotton III, Veil, & Iannarino, 2015), the impacts of the disaster in other countries (Bauer, Gylstorff, Madsen, & Mejlgaard, 2019; Kristiansen, Bonfadelli, & Kovic, 2018; Renzi, Cotton, Napolitano, & Barkemeyer, 2017), media coverage and its adequacy (Park, Wang, & Pinto, 2016; Lazic & Kaigo, 2013), communicative construction of nuclear power and nuclear risk (Kinefuchi, 2015; Kinsella, 2012), public perception and acceptance of crisis communication (Oshita, 2019), social activism and movements (Aonuma, 2019; Kinsella, 2015), and the role of popular culture in discourse ethics (Moscato, 2017).

This book joins this productive space of nuclear power communication scholarship with particular attention to the discursive formation of nuclear power. In reviewing communication scholarship on nuclear power, Kinsella, Andreas, and Endres (2015) argued that nuclear power serves as an important site of communication inquiry for understanding the relationship between materiality, language, and human agency, because its intractable materiality appears to disobey communicative construction but is nonetheless constituted by language and human agency "through the processes such as naming, constructing categories, and articulating relationships and boundaries" (p. 278). Building on this argument and following Foucault (1972) and Hall (1997), I begin with an assumption that material reality exists, but nothing meaningfully exists outside discourse. Nuclear power has a material reality – the power plants, reactors, the people who work there, utility companies, the fuels, businesses, communities that support it, inherent risks, waste, and the list goes on. But a whole series of discursive practices went into shaping nuclear power as an integral and imperial reality in Japan. Similarly, there are abundant discursive practices that have challenged that dominance to construct the meaning of nuclear power differently. My goal is to illuminate these discourses to further the understanding of the sociocultural dimensions of nuclear power. If these dimensions are communicative constructions, then, their analyses can inspire new discursive and thus material possibilities. The next section outlines the theoretical and conceptual underpinning of this book.

Discourse, hegemony, and antagonism

Michael Foucault (1972/2010) saw discourse as consisting of multiple statements that belong to the same formation. A "statement" here may mean any group of signs that have an enunciative function that includes a subject or a speaker, a referent (though the recurrence of the relation is not governed by any predetermined rules), an associative field that consists of other statements, and a material medium through which it is expressed. The multiple statements that constitute a discourse are not necessarily similar and may even be diverse but are systematically linked. Foucault (1972/2010) called this regularity *discursive formation* that occurs "[w]henever one can describe, between a number of statements, such a system of dispersion, whenever, between objects, types of statement, concepts, or thematic choices, one can define a regularity" (p. 38). For Foucault, discourse has other personalities. A discourse does not assume a single source; it may be produced by multiple subjects and in multiple contexts. However, adopting a discourse means positioning oneself with it as the subject of the discourse. Although we may not be the author of the discourse, we are nonetheless aligning ourselves with the discourse. I add here, then, that the subject also becomes a moral agent in support of the discourse. Discourse is also an open system that draws upon other discourses; it newly or differently arranges them together to create its own meanings (Hall, 1992). Such rearrangements allow room for ruptures and discontinuities, which the existing discourse tries to mend by incorporation or adaptation, but it can also lead to new articulations. Finally, discourse always involves the question and struggle of power. Power gives rise to discourse, and some discourses are more powerful than others. A powerful discourse has the ability to make things true regardless of whether or not it is based on facts (Hall, 1992). Hence, power and knowledge are implicated in each other; power produces knowledge (Foucault, 1980).

Stuart Hall (1992) used the term "discursive practice" to capture Foucault's use of "statement" but blurs the distinction between discursive and nondiscursive as do Ernesto Laclau and Chantal Mouffe (1985/2001). Hall argued that discursive practice produces meaning, and, because all social practices involve meaning, they are also discursive. Laclau and Mouffe (1985/2001) similarly argued that discourse has a material character because objects cannot be constituted as something outside discourse. Following these arguments, I see discourse as social practices that produce knowledge. Discourse encompasses thought, language, and conduct. The use of language, which may include verbal, visual, auditory, and other communicative forms, produces meaning and hence influences conduct, but the language is itself based on thought and is a product of conduct. Thus, thought, language, and conduct shape each other and shape discourse. Hall's (1997) explanation of the function of discourse summarizes why we must pay attention to discourse:

> It governs the way that a topic can be meaningfully talked about and reasoned about. It also influences how ideas are put into practice and used to regulate the conduct of others. Just as discourse "rules in" certain ways talking about a topic, defining an acceptable and intelligible

way to talk, write, or conduct oneself, so also, by definition, it "rules out," limits and restricts other ways of talking, of conducting ourselves in relation to the topic or conducting knowledge about it.

(p. 44)

Discourse hence exerts considerable power and plays a central role in shaping our identity, social relations, occurrences, experiences, and the whole culture.

Nuclear power can be examined as a site of competing discourses. Japan's nuclear power program was given birth and became an integral part of Japanese society because individuals, groups, and institutions produced statements across time and place to construct a favorable character of nuclear power. These statements may or may not be noticeably related but belong to a system of formation and may be described as a *pronuclear* discourse. Utility companies, the government, politicians, and academic experts who constitute *genshiryoku mura* (the Nuclear Village) are obvious authors and subjects of a pronuclear discourse.[1] Additionally, mass media, education, and even segments of the towns targeted for the construction of nuclear facilities have actively participated in the production of this discourse. On the other hand, nuclear power has been represented in an unfavorable light. A wide range of individuals, communities, and organizations have questioned, critiqued, and opposed nuclear power, thereby participating in the production of an *antinuclear* discourse. The main authors and subjects of this discourse include local communities where the site fights occurred, antinuclear organizations and activists, academics, public intellectuals, and mass media.

Each type of discourse, as Hall's quote given earlier clarifies, has its coherence and regulates what may or may not be said and done. However, to the extent that the producers of the pronuclear discourse represent mega-corporate interests, have enormous financial and political capital to shape discursive practices, and have the power to control the policy regarding nuclear power, it is fair to say that the pronuclear discourse has been dominant and has become hegemonic. Hegemony is inclusive of culture and ideology but go beyond them. Raymond Williams (1977) explains hegemony as "a whole body of practices and expectations, over the whole living: our sense and assignments of energy, our shaping perceptions of ourselves and our world. It is a lived system of meanings and values . . ." (p. 110). Hegemony hence appears as a constant reality for the majority of society that is, "in the strongest sense a 'culture', but a culture which has also to be seen as the lived dominance and subordination of particular classes" (Williams, 1977, p. 110). A particular segment of society, a particularity, be it a group or a perspective, comes to assume universality and dominate other groups or perspectives but the subordinated do not see the domination as coercive; they become willing or unsuspecting participants in its construction.

If the majority of the Japanese public did not think about nuclear power until the Fukushima accident, it is because this hegemony was successful. In the field dominated by pronuclear discourse, alternative narratives became unthinkable. This domination is not a historical accident but reflects motivated interests of the Nuclear Village and is made possible because power circulates through discourse and produces a regime of truth (Foucault, 1984). According to Foucault, every society has its regime of truth that it accepts as true. For industrialized capitalist society, this truth materializes "through apparatuses of diffusion and information . . . it is produced and transmitted under the control, dominant if not exclusive, of a few great political and economic apparatuses" (Foucault, 1984, p. 73). The problem that Foucault raised is not about detaching truth from power (because truth is birthed through power) but about "detaching the power of truth from the forms of hegemony, social, economic, and cultural, within which it operates at the present time" to "ascertain the possibility of constituting a new politics of truth" (Foucault, 1984, pp. 74–75).

One of the projects of this book, then, is to attend to the hegemonic pronuclear discourse to elucidate a range of social, discursive practices that have shaped the regime of truth/knowledge. I understand these practices as articulations that constitute nodal points or moments where elements gain partially fixed meaning through chains of equivalence (Laclau & Mouffe, 1985/2001). The elements are, as Laclau and Mouffe (1985/2001) insist, floating signifiers that belong to the field of discursivity and are never completely or permanently fixed. The elements that are contingently brought together have no necessary equivalence (Hall, 1996). This is why hegemony is not an end state or can never be completely achieved but is always a process that "has continually to be renewed, recreated, defended, and modified" (Williams, 1977, p. 112). For nuclear power to stay as a hegemonic reality, the members of the Nuclear Village and their supporters must constantly valorize it through articulation. This ongoing activity, past and present, is discussed in the book.

Every articulation has its excess – its constitutive outside that it excludes (Hall, 1996). We must pay attention to not only what is said or done but also what is absent and what discursive strategies of inclusion and exclusion are deployed. The idea of discursive closure is instructive here. Communication scholar Stanley Deetz (1992) wrote that discursive closure is a systematic distortion of communication and occurs whenever alternative discourses or potential conflicts are suppressed. He identified specific forms of closure – disqualification, naturalization, neutralization, topical avoidance, subjectification of experience, meaning denial and plausible deniability, legitimation, and pacification. Although Deetz's primary contextual reference is interpersonal and organizational, the idea of discursive closure is helpful in understanding how truth is accomplished in nuclear power discourses, particularly the hegemonic one and its treatment of antagonisms. Neutralization, for example, is a process of hiding the value-laden nature of a position by employing objectivity, matter-of-factness, or other rhetorical strategies (Deetz, 1992) and has been heavily utilized in pronuclear discourse as will be shown in later chapters.

Deetz saw discursive closure as a tool in the service of corporate colonization of the life world where instrumental, scientific, technical reasoning is privileged, and other forms of reasoning are marginalized. In this techno-scientific-instrumental reasoning, efficiency and effectiveness are the worthwhile criteria for the evaluation of life (Deetz, 1992). Ecofeminist philosopher Val Plumwood would agree with Deetz and add that such reasoning is constituted by and constitutive of hierarchical dualism. Her central concern is the dualized relationship between humans and nature where humans assume superiority. I will address this concern in the conclusion of the book, but here I call attention to hierarchical dualism as a principal rationality of the modern industrialized culture found at the core of hegemonic discourse of nuclear power. Plumwood (1993) argues that contrasting pairs (e.g., culture/nature, reason/nature, male/female, human/nature, reason/emotion) that are commonly deployed in discourse do not simply occupy opposite spaces; this "dualistic construction, as in hierarchy, the qualities (actual or supposed), the culture, the values and the areas of life associated with the dualized other are systematically and pervasively constructed and depicted as inferior" (p. 47). The master's identity relies on the inferiorized other, but it denies this dependency. Furthermore, once the domination takes grip of culture, the inferiorized participates in the maintenance of the inequality even though the hierarchy is not essential. Hence, Plumwood calls hierarchical dualism a logic of colonization. She introduced a number of characteristics of dualism, including backgrounding, radical exclusion, incorporation, instrumentalism, and homogenization. For both Deetz and Plumwood, the geopolitical context is the West. However, capitalism, industrialism, and corporatism have colonized much social imaginary globally, regulating and defining social relations, meanings, and values of the life world. Certainly, Japan is not outside this control. These features, as will be demonstrated in this book, populate pronuclear discourse sometimes in conjunction with discursive closure processes and other times separately.

As hegemony is power through consent, civic society must be turned into willing participants through non-coercive social control. In his study of the conflicts between civil society and the state over the construction of contentious facilities such as power plants and airports in Japan, the United States, and France, political scientist Daniel Aldrich (2008) identified social control tools the state employs to persuade its civil society. Four types of tools vary along the goal and the mode of power: coercion is used to punish resistance and ranges from physical forces to surveillance and information gathering; hard social control sets the agenda to block alternative plans or citizen mobilization; incentives are a type of soft control that induces compliance through tangible benefits; and soft social control aims to educate or socialize citizens in order to change their minds (Aldrich, 2008). Aldrich demonstrates that all of these were employed by the Japanese state, but the softer social control tools have particularly played significant roles in shaping nuclear policies in Japan, indicating the importance of civic society. I will build on the insights from this study to examine the discursive struggles rising from the use of the social control tools by the state and utilities and the ways in which these tools assume and enhance hierarchized dualism.

Although the processes of discursive closure, features of dualism, and soft control tools are deployed to fasten a discourse as the truth supported by hierarchical dualism, this final closure is not possible due to the work of antagonisms. Laclau and Mouffe (1985/2001) argue that antagonism "escapes the possibility of being apprehended through language, since language only exists as an attempt to fix that which antagonism subverts" (p. 125). Antagonisms are not external to a discourse but constitute the limit of the discourse. Thus, all discourses have their antagonisms. With regard to a hegemonic discourse, antagonisms create the possibilities of intervention through investigation, disarticulation, and rearticulation (DeLuca, 1999). Communication scholar Kevin DeLuca (1999) gives an example that the hegemonic discourse of the "American Dream" has faced antagonisms such as slavery, segregation, oppression of women, and exploitation of workers, which in turn led to struggles, namely the Civil War, the civil rights movement, the women's suffrage and liberation movements, and the labor movement. Japan's antinuclear movements, too, are a response to the hegemonic discourse's inability to achieve totality of the meaning of nuclear power. Threats to communities, lack of citizen engagement, and the very existence of the nuclear program were all antagonisms that the antinuclear activism has addressed. Thus, the second project of this book is to attend to the antagonisms as translated into antinuclear activism and movements. These antagonisms create possibilities to question the hegemonic construction of nuclear power.

The chapters

Chapter 2 charts a brief social history of Japan's nuclear power to give a general, albeit incomplete, overview of the development of the nuclear power program and the antinuclear power movements. Beginning with the atomic bombs, the chapter follows the development of two opposing paradigms of nuclear power – nuclear power as a miracle technology to boost Japan's economy and nuclear power as a technology incompatible with life. Mobilizing massive financial and political capital, the proponents aggressively built nuclear power stations, turning Japan into the third largest owner of nuclear power reactors in the world. In the meantime, the antinuclear power movements continued. The chapter is organized chronologically up to the aftermath of the Fukushima nuclear disaster to provide an overview of the ways nuclear power was both promoted and resisted in Japan in response to catalytic events.

The next two chapters give attention to the discursive formation of pro-nuclear views before the Fukushima disaster. Chapter 3 examines the role of mass media in shaping public discourse of nuclear power. Using framing as an analytical tool, the chapter examines the editorials of two leading national newspapers, *The Yomiuri Shimbun* and *The Asahi Shimbun*, from 1954 to 2000. Two questions guided the study to discern the similarities and differences between the two papers: how did the papers depict the same topics, and what topics were addressed by only one paper and how were they depicted? The two papers are commonly understood to represent contrasting political leanings. However, the analysis shows that their perspectives were not so distinct for the majority of Japan's nuclear power history. The two papers began to diverge in late 1980s with *Asahi* increasingly questioning the safety of nuclear power as a whole.

Chapter 4 covers broader aspects of the discourse that helped to hegemonize nuclear power in Japan from late 1980s until 2011. Using articulation as a method, the analysis considers a range of social practices that were employed by Nuclear Village to promote positive images of nuclear power, including children and family-friendly facilities and activities, public relations efforts by and for women, and the use of celebrities and television commercials. Each of these tools helped to construct and maintain the articulations of nuclear power as safe, necessary, and green energy, thereby weaving nuclear energy into the social, economic, and political fabrics of the Japanese society. Taken together, Chapters 3 and 4 advance the argument that pronuclear discourse by the nuclear industrial complex effectively connected nuclear energy to the well-being of Japanese society.

The pronuclear efforts were, however, not without resistance. Chapters 5 and 6 shift attention to the oppositional discourse. Pre-Fukushima antinuclear movements, particularly the local ones, are relatively unknown but are significant to the understanding of Japan's nuclear power discourse. Every community chosen for the construction of a nuclear facility experienced conflicts over whether to accept the facility. In each case, the construction project divided the community, and the struggle spanned over several decades. Some communities were successful in their antinuclear movements to reject nuclear facilities, whereas others were not. Why were some of those movements able to reject the project despite the massive financial and political power the nuclear-industrial complex held? Chapter 5 explores this question by charting both unique and common social practices that led the movements' victory in two communities.

While the local communities were the ground zero of the fight against the nuclear program, activists in urban cities were also fighting long before the Fukushima Daiichi nuclear disaster. Chapter 6 tells stories of these veteran urban activists to understand their experiences and points of view that were categorically overlooked by the media and in scholarship alike. The chapter first briefly discusses the beginning of the urban antinuclear movements that exclusively focused on nuclear power issues apart from nuclear weapons. Then, the rest of the chapter will feature seven veteran activists, addressing their motives and the themes that emerged from their interviews, including activism as identity, passive democracy as a barrier to social activism, social marginalization of antinuclear power movements, and media as a barrier to antinuclear power movements. Their stories provide broader insights into challenges of engaging in activism in Japan and antinuclear activism in particular.

The next two chapters focus on the aftermath of the Fukushima Daiichi nuclear disaster in March 2011. As the most severe nuclear power accident in Japan, it became the catalyst for the largest social movement Japan saw in almost a half century. Chapter 7 presents a variety of ways by which citizens participated in the antinuclear movement in post-Fukushima Japan. Three sites are of particular interest of the chapter: the large and regular demonstrations that took place in Tokyo, an occupy-style encampment in front of the Ministry of Economy, Trade and Industry (METI) building, and the protests that occurred in an attempt to prevent the restart of a reactor after two months of the first nuclear-free period Japan experienced since 1970. The antinuclear movements embodied at these sites articulate changing faces of Japan's social movements and their role in civic society.

Chapter 8 addresses Japan's return as a nuclear nation. The elections in December 2012 brought pronuclear Shinzo Abe and his Liberal Democratic Party (LDP) back to power. In his speech to the International Olympic Committee, pitching Tokyo as the host of the 2020 Summer Olympic, Abe declared that Fukushima was "under control." The chapter discusses the Nuclear Village's discursive practices to advance a "lessons from Fukushima" narrative in which the lessons are framed as a matter of more advanced technology to control future disasters and the theme of "successful recovery of Fukushima communities." However, ten years later, many decommissioning challenges remain, and Fukushima communities around the devastated plant continue to struggle with recovery. Hence, the chapter also attends to these antagonisms that challenge the pronuclear narrative. The counternarrative draws attention to the aspects of the post-Fukushima experiences and viewpoints left out of the Fukushima "under control" narrative.

The concluding chapter (Chapter 9) first briefly summarizes the topics covered in the preceding chapters. The rest of the chapter attends to two broader questions that emerged in the course of this project – democracy and sustainability. The question of democracy is addressed with regard to what this study of nuclear power discourses reveals about the lack of citizen participation opportunities in shaping policies that affect them. The chapter also problematizes the articulation of nuclear power as sustainable energy indispensable to the creation of a sustainable world. Unsustainability of nuclear power is discussed based on the ideas of "shadow place," ecological principles, and the logic of colonization. I end the chapter and the book by calling for an ecojustice approach to sustainability.

Note

1 "Mura" is used to emphasize the insular and conservative nature of the culture that promotes nuclear power. It is thus used as a slur to criticize the culture.

References

Aldrich, D. P. (2008). *Site fights: Divisive facilities and civil society in Japan and the West*. Ithaca, NY: Cornell University Press.

Aonuma, S. (2019). Contentious performance and/as public address: Notes on social movement rhetorics in post-Fukushima Japan. *International Journal of Communication (19328036), 13*, 3293–3320.

Arce, M., & Rice, R. (2019). The political consequences of protest. In M. Arce & R. Rice (Eds.), *Protest and democracy* (pp. 1–21). Calgary: University of Calgary Press.

Bauer, M. W., Gylstorff, S., Madsen, E. B., & Mejlgaard, N. (2019). The Fukushima accident and public perceptions about nuclear power around the globe – a challenge & response model. *Environmental Communication, 13*(4), 505–526. https://doi.org/10.1080/17524032.2018.1462225

Behling, N. H., Behling, M. C., Williams, M. C., & Managi, S. (2019). *Japan's quest for nuclear energy and the price it has paid: Accidents, consequences, and lessons learned for the global nuclear industry*. Cambridge, MA: Elsevier.

Brown, A. (2018). *Anti-nuclear protest in post-Fukushima Tokyo*. London: Routledge.

Chiavacci, D., & Obinger, J. (Eds.). (2018). *Social movements and political activism in contemporary Japan: Re-emerging from invisibility*. London: Routledge.

Cotton III, A. J., Veil, S. R., & Iannarino, N. T. (2015). Contaminated communication: TEPCO and organizational renewal at the Fukushima Daiichi Nuclear power plant. *Communication Studies, 66*(1), 27–44. https://doi.org/10.1080/105 10974.2013.811427

Deetz, S. A. (1992). *Democracy in an age of corporate colonization*. Albany, NY: State University of New York Press.

DeLuca, K. (1999). Articulation theory: A discursive grounding for rhetorical practice. *Philosophy and Rhetoric, 32*(4), 334–348.

Foucault, M. (1972/2010). *The archaeology of knowledge and the discourse on language* (A. M. Sheridan Smith, Trans.). New York: Vintage Books.

Foucault, M. (1980). *Power/knowledge: Selected interviews and other writings, 1972-1977*. New York: Pantheon.

Foucault, M. (1984). *The Foucault reader*. P. Rabinow (Ed.). New York: Pantheon.

Hall, S. (1992). The West and the rest: Discourse and power. In S. Hall & B. Gieben (Eds.), *Formations of modernity* (pp. 276–320). Cambridge: Polity Press.

Hall, S. (1996). Introduction: Why needs identity? In S. Hall & P. DuGay (Eds.), *Questions of cultural identity* (pp. 1–17). Thousand Oaks, CA: Sage Publications, Inc.

Hall, S. (1997). The work of representation. In S. Hall (Ed.), *Representation: Cultural representations and signifying practices* (pp. 15–64). Thousand Oaks, CA: Sage Publications, Inc.

Kinefuchi, E. (2015). Nuclear power for good: Articulations in Japan's nuclear power hegemony. *Communication, Culture & Critique, 8*(3), 448–465. https://doi.org/10.1111/cccr.12092

Kinsella, W. J. (2012). Environments, risks, and the limits of representation: Examples from nuclear energy and some implications of Fukushima. *Environmental Communication, 6*(2), 251–259. https://doi.org/10.1080/17524032.2012.672928

Kinsella, W. J. (2015). Rearticulating nuclear power: Energy activism and contested common sense. *Environmental Communication, 9*(3), 346–366. https://doi.org/10.1080/17524032.2014.978348

Kinsella, W. J., Andreas, D. C., & Endres, D. (2015). Communicating nuclear power: A programmatic review. *Annals of the International Communication Association, 39*(1), 277–309. https://doi.org/10.1080/23808985.2015.11679178

Kristiansen, S., Bonfadelli, H., & Kovic, M. (2018). Risk perception of nuclear energy after Fukushima: Stability and change in public opinion in Switzerland. *International Journal of Public Opinion Research, 30*(1), 24–50. https://doi.org/10.1093/ijpor/edw021

Laclau, E., & Mouffe, C. (1985/2001). *Hegemony and socialist strategy: Towards a radical democratic politics* (2nd ed.). London: Verso. (Original work published 1985).

Lazic, D., & Kaigo, M. (2013). US press coverage of the Fukushima nuclear power plant accident: Frames, sources and news domestication. *Media Asia, 40*(3), 260–273. https://doi.org/10.1080/01296612.2013.11689975

Moscato, D. (2017). Fukushima fallout in Japanese manga. *Journal of Communication Inquiry, 41*(4), 382–402. https://doi.org/10.1177/0196859917712232

Oshita, T. (2019). The effects of emergency preparedness communication on people's trust, emotions, and acceptance of a nuclear power plant. *Environmental Communication, 13*(4), 472–490. https://doi.org/10.1080/17524032.2018.1426618

Park, D. J., Wang, W., & Pinto, J. (2016). Beyond disaster and risk: Post-Fukushima nuclear news in U.S. and German press. *Communication, Culture & Critique, 9*(3), 417–437. https://doi.org/10.1111/cccr.12119

Plumwood, V. (1993). *Feminism and the mastery of nature.* London: Routledge.

Renzi, B. G., Cotton, M., Napolitano, G., & Barkemeyer, R. (2017). Rebirth, devastation and sickness: Analyzing the role of metaphor in media discourses of nuclear power. *Environmental Communication, 11*(5), 624–640. https://doi.org/10.1080/17524032.2016.1157506

Steinhoff, P. G. (2018). The uneven path of social movements in Japan. In D. Chiavacci & J. Obinger (Eds.), *Social movements and political activism in contemporary Japan: Re-emerging from invisibility* (pp. 27–50). New York: Routledge.

Tamura, A. (2018). *Post-Fukushima activism: Politics and knowledge in the age of precarity.* New York: Routledge.

Williams, R. (1977). *Marxism and literature.* Oxford: Oxford University Press.

Yamaaki, S. (2012). *Genpatsu wo tsukurasenai hitobito.* Tokyo: Iwanami Shinsho.

Zinn, H. (2007). *A power governments cannot suppress.* San Francisco: City Lights Books.

2 Japan's nuclear power
A short history

1945: Hiroshima

August 6, 1945. It was a beautiful summer morning in Hiroshima City. Not even a cloud in the sky. No one expected what was about to happen next. 8:15 a.m. An American B-29, *Enola Gay*, dropped the first atomic bomb nicknamed "Little Boy" on Hiroshima City. The bomb exploded 600 meters above the ground, instantly destroying five square miles of the city and killing 100,000 people. The ground temperature reached 4,000 degrees Celsius. By December, 40,000 more died (Yoshino, 2015). Ms. Mie Aoki (2010) was 23 years old when she was exposed to radiation from the nuclear bombing just a mile away from where the bomb was dropped. She described that day in *The Asahi Shimbun*'s series, Memories of Hiroshima and Nagasaki:

> I looked up at the blue sky from a window, thinking that I should do laundry. It was then I saw intense light million times brighter than magnesium flash by me. Then, my house began to collapse, and I balled up. A little later I regained consciousness and realized that I could move my body. . . . My younger sister who was in the next room was buried under rubble. . . . I told her, "I'm going to get help. Be strong." She nodded. I managed to get out of the house through the roof that was now on the ground. Once I was outside, I was too shocked to talk. As far as I could see, it was the sea of ruins. . . . People were walking around, moaning from pain. Their skin was hanging like rags and their faces were red and swollen like a ball. Everywhere I went, I saw dead people with their eyes open. I went back home to check on my sister. She muttered weakly, "Sister, don't worry about me. Please get out of here and save yourself." I went out to find help again. . . . When I finally grabbed a man and convinced him to help me, my house was surrounded by fire. He shouted, "It is too late. You will burn to death if you stay here. You must leave!" and he left. . . . As I walked away, I apologized to my sister over and over, begging for her forgiveness. The heat from the road was

DOI: 10.4324/9781003044222-2

unbearable. I felt as though my body was burning. But I just kept walking. I could not make sense of anything.[1]

Although she survived the nightmare and has lived to be 93 years old, Mrs. Aoki has suffered all her life from an illness caused by exposure to atomic-bomb radiation. She has often shared her story with school children and community meetings in hope that the post-WWII generations learn the atrocity of wars and nuclear weapons.

August 6, 1945. In a town near Hiroshima City, a young navy officer saw a monstrous mushroom cloud rising up in the sky of Hiroshima. He was awestruck by the power of the bomb. The next thing that came to his mind was – nuclear power is Japan's future ("Genpatsu kokka," 2011). The officer was Yasuhiro Nakasone who later became the Prime Minister in the 1980s. Dubbed as one of the founding fathers of Japan's nuclear power industry, Nakasone has told this story of witnessing the mushroom cloud and thinking of nuclear power as Japan's future in multiple interviews as well as his own autobiography. Whether the story really happened is contested, especially because it was not until the next day that U.S. President Harry Truman disclosed to the public that "Little Boy" was a nuclear bomb. Nonetheless, Nakasone's reaction to nuclear bombs and Mrs. Aoki's story represents the development of two opposing responses to nuclear power in post-WWII Japan. On the one hand, the Japanese government and the utilities saw nuclear technology as the key to rebuilding and advancing Japanese society and aggressively promoted nuclear power. On the other hand, citizens engaged in antinuclear peace activism, which coalesced into antinuclear movements.

This chapter, organized chronologically, introduces some key events to illustrate the ways nuclear power was both promoted and resisted in given sociohistorical contexts. To meaningfully delineate the development of the opposing paradigms of nuclear power over the past 60 years, I adopted James Darsey's (1990) notion of *catalytic events* – events that dramatically change exigencies and thus provide appropriate conditions for particular discourse. Catalytic events are historical (not rhetorical), are extraneous to the social movement in question, significantly help with the movement, and precede rhetorical responses that constitute a definable rhetorical era (Darsey, 1990). In short, a catalytic event inevitably and considerably changes rhetoric around a social issue. Darcey developed this methodology to discern and explain rhetorical progression of social movements, thus, setting catalytic events as major occurrences that are helpful to the movements. For the purpose of this chapter, however, I open the concept to major events that significantly influenced the development of discourses about nuclear power whether the events helped or hindered antinuclear power movements. The framework helps to make sense of the chronological development of the contrasting narratives punctuated by major events that significantly influenced subsequent responses from social actors.

1950s–1960s: atoms for peace and the birth of antinuclear power movement

The nuclear bombs dropped on Hiroshima and Nagasaki killed over 200,000 people and instantly turned the cities into ashes. Notwithstanding the catastrophe, the Japanese government saw Japan's future in nuclear technology. From the standpoint of Japan's political leadership, having been a victim of nuclear bombs was the very reason to adopt this technology. A couple of years after WWII, the young navy officer Nakasone was elected to the lower office and began his political career. He attended a summer seminar at Harvard University in 1953 and met Henry Kissinger, then a Harvard academic. Kissinger told Nakasone that the United States would be sharing nuclear technology with some countries (Johnson, 2012). Excited, Nakasone toured nuclear plants in the United States at the end of the summer seminar. In his 2006 interview with a local newspaper, Nakasone commented that his tours of the plants convinced him that nuclear power is the key to Japan's rise from its fourth-class nation status (Fukuda, 2006).

On December 8 of the same year, Dwight Eisenhower delivered his famous speech, "Atoms for Peace," at the United Nation's General Assembly. During the speech, he introduced a new vision of the United States with regard to the ongoing atomic weapon race:

> The United States knows that if the fearful trend of atomic military build-up can be reversed, this greatest of destructive forces can be developed into a great boon, for the benefit of all mankind. The United States knows that peaceful power from atomic energy is no dream of the future. That capability, already proved, is here – now – today. Who can doubt, if the entire body of the world's scientists and engineers had adequate amounts of fissionable material with which to test and develop their ideas, that this capability would rapidly be transformed into universal, efficient, and economic usage.
>
> (Press Release, "Atoms for Peace" Speech, 1953)

The speech in essence inaugurated the establishment of the International Atomic Energy Agency (IAEA) in 1957 (IAEA, n.d.).

Notwithstanding the speech, Eisenhower's call for a peaceful use of nuclear technology was not all motivated by world peace but was deeply political. In 1952, the United States successfully tested the first hydrogen bomb – the first in the world – on Eniwetok atoll in the Pacific. This gave the United States an advantage in the nuclear arms race with the Soviet Union. However, the advantage was short-lived. The Soviet Union caught up quickly when they successfully detonated a hydrogen bomb at the Semipalatinsk test site on August 12, 1953, just nine months after the U.S. testing (Arima, 2008). Washington feared that Moscow would not only be competitive but even win the nuclear arms race. Eisenhower's call for atoms for peace reflected this fear. By sharing nuclear technology with their allies and developing countries, the United States hoped that they would lead and control the development and use of nuclear technology (Arima, 2008). Promoting atoms for peace was also a solution to the enormous nuclear infrastructure the United States created through WWII. Without a war to sustain jobs, uranium enrichment facilities, nuclear scientists, and research labs would be wasted, and the country could fall into another Great Depression (Mander & Callenbach, 2011). The atoms for peace scenario provided the needed imperative to sustain the nuclear industry.

Eisenhower's UN speech served as a catalytic event that set an international ideological ground for the development of nuclear power industry. The Japanese government eagerly responded to the speech. According to Nihon Genshiryoku Sangyo Kyokai (Japan Atomic Industrial Forum, Inc.), in January 1954, the U.S. State Department sent to the Japanese government a classified document that presented nuclear power as an infinite energy source with little cost (as cited in Arima, 2008, p. 46). Following the Eisenhower's speech and this document, the Ministry of International Trade and Industry (MITI), a government agency that controlled Japan's economic and industrial policies and funding, began to consider developing a commercial nuclear power program within the framework of "Atoms for Peace." When the United Nations held the International Conference on the Peaceful Uses of Atomic Energy in Geneva in 1955, Japan sent a group of five representatives, including Nakasone, to the conference. There, the group learned about the nuclear technologies around the world and proceeded to tour nuclear power facilities in France, Germany, England, and the United States.

Convinced that Japan needed nuclear technology, Nakasone led an effort to create a political infrastructure to develop nuclear power. He and his allies introduced to the National Diet (the parliament) eight bills including the Nuclear Budget Bill of 1954 and the Atomic Energy Basic Law of 1955, which quickly passed with multi-partisan support. With the law in place, the government established the Atomic Energy Commission (AEC) within the Office of the Prime Minister as the regulatory body for nuclear power.[2] According to Nakasone, political parties, the majority and the minorities, felt the urgency to catch up with the rest of the world (Fukuda, 2006). In 1956, the National Diet approved the first nuclear power budget of 235 million yen (US$653,000 in 1956) – the number Nakasone somewhat wittily proposed based on uranium 235.

In the meantime, the citizens were not necessarily on board with the idea of "atoms for peace." The massacre and total destruction inflicted by the two bombings were a still recent memory, having occurred only a decade ago. Radiation victims were still suffering and dying. By then, the U.S. military occupation (1945–1952) had ended, and objections to nuclear technology that could not have been voiced publicly under the occupation were surfacing (Kuznick, 2011). In the midst of the rising objections came another nuclear incident. In March 1954, a Japanese fishing boat was showered with intense radiation from the U.S. thermonuclear weapon testing on Bikini Atoll. The crew of the boat, *Daigo Fukuryumaru*, was exposed to nuclear fallout from an explosion 1,000 times more powerful than Hiroshima and suffered from acute radiation syndrome. The chief radio operator died seven months later. It turned out that *Daigo Fukuryumaru* was not the only victim of the nuclear testing; the Ministry of Health and Welfare later reported that the fallout exposed 856 Japanese fishing boats and 20,000 crew members to radiation and led to dumping of 75 tons of tuna caught between that March and December (Schreiber, 2012). In addition, citizens found out that the airborne radiation from Bikini Atoll reached Japan, affecting the air they breathed and the food and drink they consumed (Arima, 2012).

This catalytic event, widely reported in the media, created uproars throughout Japan. Women's groups – housewives, mothers, teachers, and business women organizations – responded by collecting signatures against anti-bomb and antinuclear signatures (Yamamoto, 2004). By the following year, over 30 million signatures – astonishingly, one-third of the Japanese population at the time – against nuclear and hydrogen bombs were collected (Aldrich, 2013; Arima, 2008; Kuznick, 2011). Local communities, youth groups, women's organizations, and labor unions across Japan condemned nuclear weapons and nuclear testing. On August 8, 1955 – the tenth anniversary of Hiroshima – antinuclear activists held the first World Conference Against Atomic and Hydrogen Bombs. Collectively, the nationwide protests gave birth to *Gensuikyo*, the Japanese Council against Atomic and Hydrogen Bombs, and later to *Genusuikin* (the Japan Congress Against A- and H-Bombs) that objected all forms of nuclear and hydrogen technologies including civilian uses.[3] The *Daigo Fukuryumaru* accident is widely understood as the beginning of Japan's antinuclear movements (Aldrich, 2013; Higuchi, 2008; Hasegawa, 2018).

This vocal, multi-interest movement caught the attention of the U.S. government. The U.S. National Security Council suggested that the U.S. government control the Japanese protests by vigorously promoting peaceful use of nuclear power in Japan (Kuznick, 2011). In agreement, the commissioner of the AEC Thomas Murray remarked:

> Now, while the memory of Hiroshima and Nagasaki remain so vivid, construction of such a power plant in a country like Japan would be a dramatic and Christian gesture which could lift all of us far above the recollection of the carnage of those cities.
>
> ("Nihon ni genshiryoku," 1954; Kuznick, 2011)[4]

Murray and Rep. Sidney Yates (D. Illinois) introduced in 1955 a legislation to build the first nuclear power plant in Hiroshima. A few months later, the United States and Japan signed an agreement to collaborate on research and development of nuclear power (Kuznick, 2011).

The plan to build the first reactor in Hiroshima was, however, never materialized. Selling nuclear power to the Japanese public proved to be difficult in the backdrop of the nuclear bombings and the *Daigo Fukuryumaru* incident. This is when the U.S. Embassy, the U.S. Information Services, and the Central Intelligence Agency (CIA) approached Matsutaro Shoriki, the president of a national newspaper *Yomiuri Shinbun*, founder of Nippon Television Network (Japan's first commercial television station) and father of Japanese baseball (Kuznick, 2011). Shoriki had previously established a connection to CIA when he elicited assistance from them in developing a nationwide microwave communication network. For the U.S. government, Shoriki was an ideal spokesperson. Not only was he already known as an avid advocate of peaceful use of nuclear power, but he also possessed enormous power in disseminating the message of atoms for peace to the general public. Beginning January 1954, *Yomiuri Shinbun* printed a major series called *Tsuini taiyo wo toraeta* (roughly translated, "We finally captured the Sun") that introduced peaceful use of nuclear power and its benefit to society. Additionally, the newspaper frequently reported on the progress related to peaceful use of nuclear power in the United States. In November 1955, Shoriki cosponsored a U.S. exhibit of atoms for peaceful use in Tokyo. On Shoriki's end, promoting nuclear power was a means to advance his political ambition; he was planning to run for the parliament on the ticket of bringing a nuclear power industrial revolution to Japan (Arima, 2008). Shoriki won a seat to the Lower House and soon became the first chair of the Japan Atomic Energy Commission (JAEC). His newspaper continued to run stories to promote nuclear power (Arima, 2008).

The first experimental reactor was built in Tokaimura in Ibaragi Prefecture in 1963. On October 26 of the same year, Japan succeeded in the generation of electricity from nuclear power for the first time. October 26 (of 1958) was also the day Japan joined the IAEA Board. To recognize these milestones, the government designated this day as the Nuclear Power Day (The Institute for Environmental Sciences, n.d.). Beginning in 1964, the state and nuclear power institutions celebrated this day and held events to promote understanding of nuclear power. According to Aldrich (2008), establishing this day was one of the first soft control tools the Japanese state used to promote a positive image of nuclear power. In his study of the fights over controversial facilities (nuclear power plants, dams, and airports among others) in the United States, Japan, and France, Aldrich (2008) categorized the state policy instruments into coercion, hard social control, soft social control, and incentives. In the case of Japan, he found that the state mostly employed soft social control – an instrument that "seeks to change the hearts and minds of protesters" by such means as curricular integration, fairs, and award ceremonies (Aldrich, 2008, p. 64). Indeed, in addition to the national celebration of nuclear power, the government used a variety of tools from distributing posters and pamphlets in public spaces to building free energy museums and opening nuclear facilities to the public.

Following the success of the experimental reactor and with the aid of soft control and incentives, the construction of a series of commercial reactors began in the 1960s in Mihama, Tsuruga, and Fukushima, including the now crippled reactors at the Fukushima Daiichi Nuclear Station. But the push toward nuclear power was not without resistance. Throughout the 1960s, there were resilient protests from diverse groups. For example, when Mitsubishi Material built a research nuclear reactor within its research facility in Saitama City (now Oomiya City) in Saitama Prefecture in 1959, the residents protested, and 1,600 plaintiffs brought a lawsuit against the company, demanding decommissioning of the reactor (Miyazawa, n.d.). The case ended in 1974 when Mitsubishi Material agreed to close the reactor. Similarly, strong protests by local residents occurred in local communities where nuclear power plants were proposed, including Tomari-mura in Hokkaido, Hamaoka in Shizuoka, and Kashiwazaki in Niigata (Aldrich, 2008). In the 1960s to the early 1970s, academics, both students and professors, were vocal objectors to nuclear power as well. In 1965, 8,000 academics formed the national Japan Scientists' Congress (JSC) to protest pollutions and later established a committee that specifically addressed the danger of nuclear power (Aldrich, 2008).

1970–1980s: oil shock, Three Mile, and Chernobyl

In March 1970, the World Expo opened in Osaka. The theme was "Progress and Harmony for Mankind" with four sub-themes: the fuller enjoyment of life, the fuller usage of the gift of nature, the fuller application of technology in life, and better mutual understanding. Out of the sub-themes, it was the technology exhibits that commanded the most attention, and nuclear power was among the prominently displayed. As visitors entered the Expo, they were greeted with a sign that proudly informed that a nuclear reactor is helping to power the Expo (Johnson, 2012). Visitors could witness with their own eyes peaceful and remarkable application of nuclear power. Inside the Expo, they could also see a full-scale nuclear reactor mock-up created by the United Kingdom.

The Expo was indeed powered by the electricity generated at Tsuruga Power Station Unit 1, Japan's first commercial light-water reactor that just went online on March 14, 1970. This success and the publicity at the Expo marked the beginning of golden decades for the nuclear-industrial complex. The majority of today's existing nuclear reactors were built in the 1970s and the 1980s. This rapid development was made possible by a combination of circumstantial and strategic factors. First, the larger political circumstance made it easy for the government to promote nuclear energy. In 1973, the Organization of Arab Petroleum Exporting Countries (OAPEC) placed severe restrictions on oil export in protest to the U.S. military support for Israel in its war (Yom Kippur War) with Egypt and Syria. This caused a sharp spike in the oil prices from $3 a barrel to $12, resulting in many short-term and long-term economic and political consequences throughout the world. As a country with scarce energy resources and dependent on imported oil, Japan acutely felt the weight of this international political and economic crisis.

For the nuclear-industrial complex, the oil crisis proved to be a catalytic event – an event that provided a claim base for nuclear power. The answer to the oil shock did not have to be nuclear power; the crisis could have been an opportunity to invest in research and development of other energy sources such as wind, solar, fuel cells, and natural gas, but the government promoted nuclear power as *the* way to solve Japan's dependency on oil (Takagi, 2000). Jinzaburo Takagi, late nuclear chemist and founder of the Citizen's Nuclear Information Center (CNIC), argued that Japan's heavy dependence on nuclear power is not because of nuclear power's brilliance but because the state supported it as a matter of national policy and aggressively built nuclear power plants (Takagi, 2000). And the oil shock gave the state an opportune rationale and tangible rhetorical backing for the need to implement the policy.

The majority of the reactors (36 of the 54) that were operational at the time of the Fukushima disaster were completed and went online in the 1970s and the 1980s. What propelled this rapid construction boom? The oil crisis and the government's framing of nuclear power as the sole solution to the crisis set a larger socioeconomic tone in favor of nuclear power. However, convincing the locals to accept the construction of nuclear reactors in their backyards was another matter. The *kaku alerugi* or nuclear allergy (dislike and phobia of nuclear technology) continued to exist among citizens, given the horrific history of Hiroshima and Nagasaki and the *Daigo Fukuryu Maru* incident. The communities sited for nuclear power plants, most of which relied on fisheries and farming, were afraid that nuclear power might not only bring direct health risks in the case of an accident but also threaten their livelihood by contaminating water and farms even in its normal operation (Aldrich, 2008).

In his research of Japan's nuclear plant siting, Aldrich (2008) shows that the strength of civic society was a determining factor of where plant construction was attempted. Geographical conditions such as the absence of fault lines and accessibility to plentiful water for cooling reactors are imperative to the operation of nuclear power plants. Beyond these criteria, authorities targeted communities with diminishing levels of social capital. They were well aware that communities are less likely to fight nuclear power when their social ties are weak due to depopulation, declining fishing and farming cooperatives, or sudden population increase. Once "ideal" communities with weak social capital were identified, the government gained consent of the town officials by promising lump sums of subsidies and persuaded key players such as landowners and fishing cooperatives to sell their lands and to transfer fishing rights to the utilities in exchange for generous financial compensation. Power companies succeeded in gaining consent from many communities they targeted, and, as a result, the majority of today's reactors were built during the 1970s and 1980s.

The success of the pronuclear bloc, however, does not mean that the antinuclear movement birthed in the previous decades was dead. On the contrary, there were notable developments on that front. Among the major local antinuclear movements were the series of protests that occurred in Niigata Prefecture regarding the construction of the Kashiwazaki-Kariwa Nuclear Power Plant by Tokyo Electric Power Company (TEPCO). In 1972, the residents of Kashiwazaki City and Kariwa Village opposing nuclear power demanded a referendum to annul a plan to construct a nuclear power plant in the area shared by the two communities. After years of conflict, TEPCO and the central government attempted a joint public hearing in 1980 to appease the residents. However, 6,000 protesters comprised of the residents, members of *Gensuikin*, labor groups, and the Socialist Party gathered to object the plant construction (Aldrich, 2008). The government responded to the protest with 2,000 riot police to guard the building where the hearing took place. Despite the large-scale and persistent objection over 17 years, the plant construction proceeded and was completed in 1987. Still, the struggle in Kashiwazaki-Kariwa illustrates that antinuclear movements were far from defunct.[5] It is also a testament that communities did not simply welcome nuclear power in their backyard; each of the communities that eventually accepted nuclear power experienced various degrees and periods of intense fights and conflicts.

Citizens' protests were not limited to siting and construction. In Mihama and Ōi, local citizens and many members of the Union of Electric Utility Workers – some of whom were employed at nuclear power plants – denounced the lack of transparency and poor safety records at the plant (Aldrich, 2008). In response, the MITI opened local branches where citizens could directly talk to local government representatives. Although these offices rarely provided information regarding accidents, radioactive waste storage, radiation's health risk, or other information that citizens are likely to find objectionable (Aldrich, 2008), at minimum, the opening of the MITI branches demonstrates that citizens do and did have the power to pressure the MITI to respond to their protests.

In addition to local protests and efforts to pressure nuclear power authorities, key national antinuclear power organizations emerged in the 1970s as well. In 1975, Citizens' Nuclear Information Center (CNIC) was established by nuclear chemist Jinzaboro Takagi. With the goal of realizing nuclear power-free society, CNIC has conducted research and collected information related to nuclear power and has shared the information with the public. It remains today as an important information hub and watchdog of nuclear power in Japan. *Han genpatsu undo zenkoku renraku kai* (the National Liaison Conference of the Anti-Nuclear Movement) also emerged in the 1970s. Since its inception, it has published *Hangenpatsu Shimbun*, an antinuclear newspaper that seeks to connect local movements, share local

Figure 2.1 (Left) An old dilapidated building (with "Protect grandchildren's future" written on the side) that served as the headquarter of Kariwa's antinuclear movement. (Right) State-of-the-art Kariwa Village Lifelong Learning Center RAPIKA built with nuclear power subsidies.

Source: Photograph by the author.

developments, and provide energy-related information. In 1987, *Genpatsu mondai jūmin undō zenkoku renraku sentah* (Genjūren), a national network of citizens' antinuclear movement, was launched to address nuclear power safety, to connect citizens across communities, and to call for the elimination of mixed oxide (MOX) fuel and overall nuclear-free society (Genjūren, n.d.).

Antinuclear movements grew in the 1970s and 1980s also because two of the worst nuclear disasters in the world occurred during these years. On March 26, 1979, a cooling malfunction occurred at the No. 2 reactor of the Three Mile Island (TMI) Nuclear Generating Station in Pennsylvania, which led to partial meltdown of the reactor's core. This most serious commercial nuclear accident in the United States to date was given level-five designation on the International Nuclear and Radiological Event Scale (INES) that ranges from zero to seven. The U.S. Nuclear Regulatory Commission and World Nuclear Association (2020) concluded that there were no adverse health effects from the nuclear accident. However, conflicting stories were reported by NGOs and investigative journalists. For example, in *Voices from Three Mile Island*, a two-hour public radio documentary that aired a year after the accident, independent filmmaker Robbie Leppzer represented the voices of the residents who lived within five-mile radius of the plant. The residents expressed fear, anger, and frustration, including local farmers who reported grave health problems among farm animals such as birth defects, stillborn, and miscarriages (Leppzer, n.d.). Journalists Harvey Wasserman and Norman Solomon (1982) found from over 200 interviews with residents living near the plant that they had abnormally high rates of infant deaths and stillbirth, birth defects, disappearance of animals and insects, unexplained deaths, wilting of vegetation following the nuclear accident. *Three Mile Island: The people's testament* (Smith, 1989), based on over 250 interviews with residents over several years, reported similar findings. These stories made their ways to Japan to present viewpoints and experiences unrepresented by the U.S. government.

Just seven years after TMI, another even larger nuclear catastrophe shook the world. On April 26, 1986, what was supposed to be a safety enhancing experiment at the Chernobyl Nuclear Power Plant in Ukraine (then a part of the Soviet Union) turned into the worst nuclear power disaster in world history. Engineers were testing to see if the cooling pump system would work using the electricity generated from the reactor in the event the secondary electricity supply failed. The engineers lowered control rods to reduce the output for the experiment, but too many rods were lowered, and output dropped too quickly. The engineers raised some rods to increase outputs, but power surge suddenly occurred, and the reactor began to overheat. The emergency shutdown button was pushed, but the attempt to lower the heat by reinserting the removed control rods failed. With extreme overheat, the fuel pellets in the core exploded, blowing off the reactor's roof. Radioactive contents began to spill. The air sucked into the reactor ignited fire, which subsequently burned for nine days. Massive amounts of radioactive materials as much as 400 times more powerful than Hiroshima were released into the atmosphere (The International Atomic Energy Agency, 1996).

These two nuclear accidents shocked Japan's civic society at the time of the accidents. TMI raised awareness among the communities near nuclear power about the dangers of nuclear power, prompting the residents to demand utilities to increase safety measures (Pickett, 2002). Chernobyl, in particular, was a catalytic event that brought much awareness to the danger of radiation; although radiation debris from the disaster did not directly fall on Japan, food imported from Europe was found to be radioactive, raising fear about the danger of nuclear power (Aldrich, 2008). Citizens nationwide rose, calling for phase-out of nuclear power. In 1988, 20,000 protesters gathered in Hibiya Park, Tokyo, for a rally organized by the CNIC. In the same year, 3.6 million people signed a petition seeking discontinuation of nuclear power program (Smith, 2011). The citizens submitted the signatures to the Diet, but "as if they were cardboard boxes of meaningless shredded paper for recycling, these signatures didn't make it very far. The Diet's Trade and Industry Committee refused to even meet with us" (Smith, 2011, p. viii). Then Prime Minister – and a founding father of nuclear power in Japan – Yasuhiro Nakasone declared that an accident like the one occurred in Chernobyl would never happen in Japan, because the reactors used in Japan are of a completely different model (Genpatsu kokka, 2011).

Although antinuclear power activists and citizen protesters faced stone-walling and non-response from the government, it did not mean that their voices had no effect. The large-scale protests prompted by Chernobyl taught the nuclear-industrial complex profound impact foreign accidents can have on Japan's nuclear power development (Pickett, 2002). As citizens became more educated about the dangers of nuclear technology and as the grass-roots efforts to subvert nuclear power grew, so did the pressure on the government to increase their effort to fortify their narrative of nuclear as safe and good. In 1988, the MITI opened a public relations unit to advertise existing nuclear power safety (Aldrich, 2008).

1990s–2011: Nuclear power nation and continuing resistance

As the government escalated the campaign for nuclear power, so did the opponents. The end of the 1980s gave birth to more antinuclear organizations promoted by the Chernobyl disaster and other domestic accidents.[6] In Tokyo, the concerns over radioactive food from Chernobyl prompted a group of citizens to test radiation in food. In 1989, this group became Tampoposha, a non-governmental organization (NGO) and began to seek phase-out of nuclear power. Similarly, Green Action was established in Kyoto in 1991 by Aileen Mioko Smith, who had previously worked to expose the suffering of *Minamana* victims as well as the voices of the TMI residents. Green Action distributes information related to new developments in the nuclear power industry and the government, organizes actions, and connects citizens with experts and domestic issues with international issues (Green Action, n.d.). Despite the scarcity of funding, government grants in particular, these and other antinuclear NGOs have been in existence for decades, resiliently leading and supporting antinuclear activism in Japan.

An important catalyst of the resilience is nuclear accidents. If TMI and Chernobyl raised public awareness about the danger of nuclear power, the 1990s brought this danger home. One of such significant accidents occurred in 1995. On December 8, 1995, Monju, a sodium-cooled fast-breeder nuclear reactor in Tsuruga City, Fukui Prefecture, experienced "criticality" or "an uncontrolled nuclear chain reaction" (World Nuclear Association, 2013). Several hundred kilograms of sodium leaked and reacted with the oxygen and moisture, causing the temperature to reach several hundred degrees Celsius. The extreme temperature warped the steel structure of the room. The leaked sodium – as much as three tons – was determined as not radioactive because the accident occurred in the secondary cooling system. Japan Atomic Energy Agency (JAEA), a government-supported agency that operated the reactor, covered up the severity of the accident and damage by falsifying reports, editing the videotape of the accident, and prohibiting employees from sharing information with the public. The cover-up, but not necessarily the accident itself, led to public outcry.

Since its outset, Monju has been a site of intense contestation. The reactor was designed to use plutonium-uranium MOX fuel made from plutonium and uranium extracted from spent nuclear fuel. It was hailed by its supporters as a "dream reactor" that would save Japan from the necessity to import uranium (Aoki, 2015). However, opponents have strongly objected to the use of MOX fuel, arguing that it is more prone to causing nuclear accidents and that it releases much higher levels of radioactive materials if an accident occurs. After the 1995 accident, the reactor remained idle for 15 years and came back online in 2010. Just three months later, it was shut down again when a fuel-loading device fell into the reactor vessel. The reactor was banned from operation under the stricter post-Fukushima safety regulations and was put on the decommission list in 2016 without recovering any of the scandalous cost. During its 22 years of life, the reactor operated only 205 days, but Japan spent over 1 trillion yen (US$12.5 billion) for construction, 50 million yen daily for maintenance even without operation, and will cost US$3.2 billion for decommissioning (World Nuclear Association, 2021). Antinuclear NGOs have exposed much of these problems occurring at Monju.

The 1990s brought a more criticality accident at a uranium-reprocessing plant operated by Japan Nuclear Fuel Conversion Co. (JCO) in Tokaimura, Ibaraki Prefecture. On September 30, 1999, three workers were preparing a small batch of fuel for a fast-breeder reactor. The fuel preparation procedure approved by regulatory authorities required specific series of steps – first, dissolve uranium oxide powder in nitric acid in a dissolution tank, second, transfer the content to a storage column for mixing with pure uranyl nitrate (a water-soluble yellow uranium salt) solution, and finally transfer the mixture to a precipitation tank covered in a water cooing jacket that removes excess heat from chemical reaction (World Nuclear Association, 2013). A few years before the accident, JCO simplified its procedure without permission from the regulatory authorities; workers were allowed to dissolve uranium oxide in a steel bucket instead of the dissolution tank. The operators further sped up the procedure by putting the solution directly into the precipitation tank, failing to precisely measure the amount of the solution and to use the storage column for mixing. As a result, too much uranium was poured into the tank, and fission products began to be released in the building. In Tokai's fuel preparation, which was a wet process, the water accelerated and prolonged the chemical reaction to last about 20 hours. The three technicians received high doses of gamma and neutron radiation, and two of them died several months later.

Chernobyl, Monju, Tokai, and other minor but frequent accidents increased citizens' awareness about real and potential dangers of nuclear power. The cover-ups of Monju and other accidents exacerbated citizen's distrust of the government and the nuclear power industry.[7] Tokai showed slow and agonizing deaths that radiation is capable of inflicting. Responding to the increasing opposition, the nuclear-industrial complex intensified their effort to win over the public. In the mid-1990s, the Agency for Natural Resources and Energy (ANRE), a METI agency, began to air pronuclear television commercials, and quasi-governmental organizations such as the Japan Atomic Energy Relations Organization (JAERO) and the Center for the Development of Power Supply Regions enhanced its public relations effort (Aldrich, 2008). As will be shown in Chapter 4, nuclear power plants welcomed visitors to their public relations facilities that resembled amusement and interactive parks to attract families and school children. Positive images of nuclear power were also embedded in school curriculums as part of science education. These soft control instruments (Aldrich, 2008) aimed at inserting nuclear power as natural and positive part of civic society.

Part of the public relations effort focused on the nuclear power's role in fighting climate change. The Earth Summit of 1992 and the Kyoto Protocol of 1997 defined climate change as a grave global problem that each state must work to alleviate. The nuclear-industrial complex seized this emergent problem as an opportunity to promote nuclear power in a new light – nuclear as a green, environmentally friendly energy unlike fossil fuels. Since the 1990s, this narrative, as will be discussed further in Chapter 4, has been promoted in television commercials, newspapers, posters in train stations, and many other places, informing and educating the public about the goodness of nuclear power.

While controlling the symbolic representation of nuclear power, the government amplified financial incentives to persuade communities of the benefits of housing a nuclear power plant. For the 1997 budget, for example, the MITI asked the National Diet for 5.09 billion yen to be put into long-term subsidy programs for these communities (Aldrich, 2008).[8] The property tax of the facilities was substantial as well. For example, a town with a 1,350 mwh nuclear power station received about 63 million yen (US$5.7 million) in property tax the first year of operation (Ito, 2011). Other financial incentives included employment and businesses created by hosting nuclear facilities. In towns with limited sources of employment and decreasing populations, subsidies and tax revenues were irresistible incentives. However, the introduction of nuclear power also severely undermined agriculture, fisheries, and other local land-based economies, creating nuclear dependency.

Despite the series of accidents and exposure of cover-ups of accidents, nuclear power continued to build its status as a dominant industry in the 1990s and the first decade of the 21st century. Several new reactors were under construction. The increasing pressure to curb carbon emission was working in favor of nuclear power. By 2010, 30% of Japan's electricity came from nuclear power, and the government planned to increase the nuclear share to 50% by 2030 (World Nuclear Association, 2010). This plan, however, came to a sudden halt on March 11, 2011, when the largest catalytic event struck Japan.

3/11: Fukushima and aftermath

On March 11, 2011, a magnitude-9 earthquake hit the Tohoku region of Japan, triggering massive 130 foot tsunami. The earthquake-tsunami caused enormous damage to the region and claimed the lives of over 18,000 people. But the tsunami also led to an unexpected, critical incident at the Fukushima Daiichi Nuclear Power Station operated by the TEPCO. Three operating reactors lost power when the station was flooded by the tsunami, which then disabled the generators and the heat exchangers that would dump waste heat to the ocean. As a result, the reactors lost proper cooling and water circulation functions (World Nuclear Association, 2016). The diesel generators, electrical switchgears, and batteries were all located in the basements and became submerged by the flood water. Without the cooling functions, the reactors began to overheat and eventually led to hydrogen explosions at all three reactors. The reactors' cores largely melted and fell into the containment vessels. While the Japanese government initially downplayed the disaster, it later determined that the severity of the disaster reached level seven, the highest on the INES. Over 160,000 residents were forced to evacuate their homes.

In 2017, the government began to lift the evacuation orders for the municipalities that were once off-limits due to the nuclear disaster. In spring of 2021, the return rate in ten municipalities near the devastated nuclear power plant remains 20–30% ("Fukushima struggling," 2021). In the meantime, the decommissioning of the plant has been slow. One of the major challenges is containment of the contaminated water. TEPCO planned to stop all water flow in and from the Fukushima Daiichi Nuclear Plant by 2020, but water continues to leak (World Nuclear News, 2021).

The enormity of the Fukushima disaster prompted large-scale antinuclear power demonstrations unprecedented in the last half century. Demonstrations began to appear weeks after the disaster. In early April, for example, there were two demonstrations in Tokyo that drew over 17,000 (Mahr, 2011). Larger and more frequent demonstrations began to appear six months after the disaster as citizens became increasingly impatient with the lack of transparency and the inability of the government to make any meaningful progress in dealing with the aftermath of the disaster. In particular, large-scale rallies occurred in Tokyo. In September 2011, over 60,000 people gathered in central Tokyo, calling for nuclear-free Japan. Around this time, in Tokyo, Osaka, and Kyoto, protesters began to gather in front of headquarters of electric power companies for demonstration every

Friday. In Tokyo, protesters began to gather in front of the Prime Minister's residence on Friday evenings, demanding phase-out of nuclear power. In June 2012, one of these Friday demonstrations at the Prime Minister's residence drew over 45,000 protesters. A few weeks later, another even larger rally occurred in Yoyogi Park, Tokyo, attracting, in the organizers' estimate, 170,000 people (Tabuchi, 2012). Mainstream news media, both domestic and international, reported the spectacles as a rare display of public dissent in Japan. Indeed, these massive demonstrations are the largest public protests in Japan in the last 50 years (Williamson, 2012).

In addition to these typical forms of protest, another form of protest emerged on September 11, exactly six months after the disaster. A group of antinuclear power activists set up *Hangenpatsu Tento* (an antinuclear tent) in front of the METI building in the middle of Tokyo. Among the policy areas that METI has jurisdiction over is energy security. As nuclear power plays *the* key role of Japan's energy independence, METI provides political infrastructure for the nuclear power industry. *Hangenpatsu Tento* in front of this ministry building is thus the most conspicuous, straightforward symbol of post-Fukushima antinuclear movement. The government sued the tent occupants for unlawfully seizing the national land, but the occupants have argued that the tent is their freedom of speech, their constitutional right. In August 2016, Tokyo District Court ruled that the occupation is illegal and forced immediate removal of the camp. The tents are gone, but activists continue to gather there and carry on a daily demonstration as of 2021.

Numerous demonstrations have also occurred near nuclear power plants. Every time a reactor was announced for restart after Fukushima, citizens gathered to protest against it. When No. 3 reactor at the Ōi Nuclear Power Station in Fukui Prefecture was back online in July 2012, the locals and people from neighboring towns and prefectures concentrated their effort to stop the restart. When No. 1 and No. 2 reactors at the Sendai Nuclear Power Station were restated in August and November 2015 respectively under new safety regulations set by the newly formed Nuclear Regulatory Authority, protestors poured into the town of Satsumasendai.[9] Although the protests were unable to stop the restarts, residents filed a lawsuit to shut down the reactors albeit unsuccessfully.

Meanwhile, another lawsuit led to a successful shutdown of two reactors in Fukui Prefecture temporarily. No. 3 reactor at the Takahama Nuclear Station went online in January 2016, and No. 4 reactor was scheduled for restart but was delayed due to technical problems. A group of residents from Shiga Prefecture filed a lawsuit, stating that an accident, if occurred at the station less than 30 km from the prefecture, would impact Lake Biwa that provides water for 14 million people in the Kansai area (Johnson, 2016). The Otsu District Court issued a provisional ruling to shut down the plant, stating remaining concerns over the tsunami response and the evacuation plan by the Kansai Electric Power Company (KEPCO) that operates the plant. This was the first injunction to halt nuclear reactors reactivated under the Nuclear Regulation Authority's (NRA's) safety regulations. A year later, however, Osaka High Court overturned the injunction, and the two reactors came back online. In 2021, nine reactors are operational.

From Hiroshima and Nagasaki to Fukushima

This chapter charted Japan's nuclear power history. It began with Japan's experience with nuclear bombs, a catastrophe that paradoxically gave birth to two opposing developments. On the one hand, it prompted a peace movement that opposed wars and the use of nuclear warheads. Peace activists refused to accept Dwight Eisenhower's rhetoric of "atoms for peace" and opposed all nuclear technologies as antithetical to peace, including nuclear power for civilian use. In the last half century, the antinuclear movement was supported and advanced independently and collectively by diverse groups – local and urban residents, members of political parties, housewives, and intellectuals and professionals (Cavasin, 2008) – at different localities and through different time periods. On the other hand, there was a contrasting development; with the advice of the United States, the Japanese government and corporate interests articulated the nuclear bombing as the evidence of the immense potential that nuclear power held in fueling the reconstruction of Japan's economy. Mobilizing massive financial and political capital, proponents aggressively built nuclear power stations, turning Japan into the third largest owner of nuclear power reactors in the world.

The predominance of nuclear power may define Japan as a pronuclear power nation where citizens by and large accepted and supported nuclear power. Indeed, 17 local communities consented to the construction of nuclear power plants and became the hosts of 54 nuclear power reactors. It is, however, important to note that dozens more proposed reactors were never built because of strong oppositions. Equally important is the forgotten history of intense conflicts that towns experienced even if they eventually came to host nuclear power reactors. Antinuclear activists and organizations have also acted as watchdogs of the nuclear power industry and the government, calling out cover-ups and bringing to the public's attention the danger of nuclear power. They have provided a counter-narrative to the narrative of nuclear power circulated widely by the nuclear-industrial complex. The following chapters will take a closer look at the social practices by which nuclear power has been both promoted and opposed.

Notes

1 An excerpt from Mie Aoki's story. The excerpt is a translation by the author.
2 The AEC included Hideki Yukawa, theoretical physicist who became the first Japanese Nobel laureate.

3 *Gensuikyo* consisted of various political and civil interest groups that were united under the banner of antinuclear weapons. In 1965, the Socialist-leaning groups broke off of *Gensuikyo* and formed *Genusuikin* (the Japan Congress Against A- and H-Bombs). Since its birth, *Gensuikin* objected all forms of nuclear and hydrogen technologies and has spoken against nuclear weapons and nuclear energy.

4 The English-language quote was cited by Kuznick. A Japanese translation of the same quote was published in *The Asahi Shimbun*.

5 The activism continued after the construction of the plant. In 2001, the residents successfully blocked through a public referendum of TEPCO's plan to introduce the controversial MOX fuel to the plant.

6 Although small, there had been many instances of accidents and malfunctions at nuclear power plants. *Gensuikin* reported 43 accidents in the 1970s, 77 in the 1980s, and 40 between 1990 and 1996.

7 In 2002, Tokyo Electric Power Company (TEPCO) was exposed for its falsification of safety data from the 1980s to the 1990s. More details at www.fepc.or.jp/nuclear/safety/past/tokyodenryoku/

8 US$47 million at a 1996 rate.

9 The Nuclear Regulation Authority (NRA) was established in September 2012 as part of the Ministry of the Environment. Before the Fukushima disaster, Nuclear Safety Commission (NSA) under the cabinet oversaw nuclear safety, reviewing inspections conducted by the Nuclear and Industrial Safety Agency (NISA), a regulatory and oversight agency housed in the METI. After the Fukushima disaster, NISA and NSA came under severe criticisms for inadequate safety administration and were replaced with the NRA.

References

Aldrich, D. P. (2008). *Site fights: Divisive facilities and civic society in Japan and the West*. Ithaca, NY: Cornel University Press.

Aldrich, D. P. (2013). Antinuclear power movement in Japan. In D. A. Show, D. Della Porta, B. Klandermans, & D. McAdam (Eds.), *The Wiley-Blackwell encyclopedia of social and political movement*. Hoboken, NJ: John Wiley & Sons. Retrieved from https://onlinelibrary.wiley.com/doi/book/10.1002/9780470674871

Aoki, M. (2010). Memories of Hiroshima and Nagasaki. *The Asahi Shimbun*. Retrieved from www.asahi.com/hibakusha/hiroshima/h00-00367j.html

Aoki, M. (2015, November 23). Fate of troubled Monju reactor hangs in the balance. *The Japan Times*. Retrieved from www.japantimes.co.jp/news/2015/11/23/reference/fate-of-troubled-monju-reactor-hangs-in-the-balance/#.VseFOMdiClM

Arima, T. (2008). *Genpatsu, Shoriki, CIA*. Tokyo: Shincho Shinsho.

Arima, T. (2012). *Genpatsu to genbaku*. Tokyo: Bunshun Shinsho.

Cavasin, N. (2008). Citizen activism and the nuclear industry in Japan: After the Tokai Village disaster. In P. P. Karan & U. Suganuma (Eds.), *Local environmental movements: A comparative study of the United States and Japan* (pp. 65–74). Lexington, KY: The University Press of Kentucky.

Darsey, J. (1990). From "gay is good" to the scourge of AIDS: The evolution of gay liberation rhetoric, 1977–1990. *Communication Studies*, 42(1), 43–66. http://dx.doi.org/10.1080/10510979109368320

Fukuda, S. (2006, March 20). Nakasone Yasuhiro moto shushou ni kiku [Interview with former Prime Minister Yasuhiko Nakasone]. *Touou Nippo*. Retrieved from www.toonippo.co.jp/rensai/ren2006/nakasone/0320.html

Fukushima struggling to get people back, 10 years after disaster. (2021, February 22). *The Japan Times*. Retrieved from www.japantimes.co.jp/news/2021/02/22/national/fukushima-residents-struggle/

Genjūren. (n.d.). Retrieved from http://homepage2.nifty.com/gjc/

Genpatsu kokka: Nakasone Yasuhiro hen [Nuclear power nation: Nakasone Yasuhiro issue]. (2011, July 21). *The Asahi Shimbun*. Retrieved from http://blog.goo.ne.jp/harumi-s_2005/e/894e9105d199310c1457db0eb7605b73

Green Action. (n.d.). Retrieved from http://greenaction-japan.org/en/

Hasegawa, K. (2018). Continuities and discontinuities of Japan's political activism before and after the Fukushima disaster. In D. Chiavacci & J. Obinger (Eds.), *Social movements and political activism in contemporary Japan: Re-emerging from invisibility* (pp. 115–136). New York: Routledge.

Higuchi, T. (2008). An environmental origin of antinuclear activism in Japan 1954–1963. *Peace and Change, 33*(3), 333–367. Retrieved from https://doi.org/10.1111/j.1468-0130.2008.00502.x

The Institute for Environmental Sciences. (n.d.). Retrieved from www.ies.or.jp/publicity_j/mini_hyakka/05/mini05.html

The International Atomic Energy Agency. (1996). *Ten years after Chernobyl: What do we really know?* Retrieved from https://inis.iaea.org/collection/NCLCollectionStore/_Public/28/058/28058918.pdf

The International Atomic Energy Agency. (n.d.). *History*. Retrieved from www.iaea.org/about/history

Ito, H. (2011, August 1). Chiki shigen wo ikashita machi zukuri e [Towards community building based on the local resources]. *Japan Agricultural Communications*. Retrieved from www.jacom.or.jp/archive03/tokusyu/2011/tokusyu110801-14409.html

Johnson, E. (2012). From Hiroshima to Fukushima: The history of nuclear power development in Japan. In *Fresh currents*. Kyoto, Japan: Kyoto Journal.

Johnson, E. (2016, March 9). *Court issues surprise injunction to halt Takahama nuclear reactors*. Retrieved from www.japantimes.co.jp/news/2016/03/09/national/court-issues-surprise-injunction-halt-takahama-nuclear-reactors/#.Vwp3P8enUkg

Kuznick, P. (2011, April 13). Japan's nuclear history in perspective: Eisenhower and atoms for war and peace. *Bulletin of the Atomic Scientists*. Retrieved from http://thebulletin.org/japans-nuclear-history-perspective-eisenhower-and-atoms-war-and-peace-0

Leppzer, R. (n.d.). *Voices from Three Mile Island*. Retrieved from www.turningtide.com/voicesfromtmi-listennow.htm

Mahr, K. (2011, April 11). What does Fukushima's level 7 status mean? *Times*. Retrieved from http://science.time.com/2011/04/11/what-does-fukushima's-new-"level-7"-status-mean/

Mander, J., & Callenbach, E. (2011). Prologue: False solutions. In E. Callenbach (Ed.), *Nuclear roulette* (pp. iv–vii). San Francisco: The International Forum on Globalization.

Miyazawa, Y. (n.d.). *Ganshiryoku sosho wo tanto shite [Experience of working on a nuclear power lawsuit]*. Retrieved from https://saitamasogo.jp/archives/68

Nihon ni genshiryoku hatsudensho wo [The proposal to build nuclear power plants in Japan]. (1954, September 22). *The Asahi Shimbun*, p. 2.

Pickett, S. (2002). Japan's nuclear energy policy. *Energy Policy, 30*, 1337–1355.

Press Release, "Atoms for Peace" speech. (1953, December 8). Retrieved from www.eisenhower.archives.gov/research/online_documents/atoms_for_peace.html

Schreiber, M. (2012, March 18). Lucky Dragon's lethal catch. *The Japan Times*. Retrieved from www.japantimes.co.jp/life/2012/03/18/general/lucky-dragons-lethal-catch/

Smith, A. M. (1989). *Three Mile Island: The people's testament*. Retrieved from www.beyondnuclear.org/tmi-truth/

Smith, A. M. (2011). Fukushima: Notes of a Japanese anti-nuclear activist. In *Nuclear roulette* (pp. viii–xii). The San Francisco: International Forum on Globalization.

Tabuchi, H. (2012, July 19). Tokyo rally is biggest yet to oppose nuclear plan. *New York Times*. Retrieved from www.nytimes.com/2012/07/17/world/asia/thousands-gather-in-tokyo-to-protest-nuclear-restart.html?_r=0

Takagi, J. (2000). *Genshiryoku shinwa kara no kaiho [de-mystifying nuclear power myths]*. Tokyo: Kodansha.

The U.S. Nuclear Regulatory Commission. *Backgrounder on the Three Mile Island accident*. Retrieved from www.nrc.gov/reading-rm/doc-collections/fact-sheets/3mile-isle.html

Wasserman, H., & Solomon, N. (1982). *Killing our own: The disaster of America's experience with atomic radiation*. New York: Dell Publishing.

Williamson, P. (2012, July 2). Largest demonstrations in half a century protest the restart of Japanese nuclear power plants. *The Asia-Pacific Journal, 10*(27). http://apjjf.org/2012/10/27/Piers-Williamson/3787/article.html

World Nuclear Association. (2010). *Nuclear power in Japan*. Retrieved from http://world-nuclear.org/info/default.aspx?id=344&terms=Japan

World Nuclear Association. (2013). *Tokaimura criticality accident 1999*. Retrieved from http://world-nuclear.org/information-library/safety-and-security/safety-of-plants/tokaimura-criticality-accident.aspx

World Nuclear Association. (2016, February 25). *Fukushima accident*. Retrieved from www.world-nuclear.org/information-library/safety-and-security/safety-of-plants/fukushima-accident.aspx

World Nuclear Association. (2020, March). *Three Mile Island accident*. Retrieved from http://world-nuclear.org/information-library/safety-and-security/safety-of-plants/three-mile-island-accident.aspx

World Nuclear Association. (2021, January). *Japan's nuclear fuel cycle*. Retrieved from www.world-nuclear.org/focus/fukushima-daiichi-accident/japan-nuclear-fuel-cycle.aspx

World Nuclear News. (2021, February 22). *Increased rate of water leakage at Fukushima reactors*. Retrieved from www.world-nuclear-news.org/Articles/Increased-rate-of-water-leakage-at-Fukushima-react

Yamamoto, M. (2004). *Grassroots pacifism in post-war Japan: The rebirth of a nation*. New York: Taylor & Francis.

Yoshino, T. (2015, August 6). Sengo 70 nen [70 years after the war]. *Huffington Post Japan*. Retrieved from www.huffingtonpost.jp/2015/08/04/news-august-6_n_7930770.html#link02

3 Mediating nuclear power for citizens

Newspaper editorials in shaping nuclear power

Newspaper editorials as framers

The mass media plays a powerful role in shaping social issues because of its extensive reach to the public. In particular, print media has long served to provide "unified fields of exchange and communication" to shape national consciousness (Anderson, 1991, p. 44). This was particularly true before the rise of the World Wide Web in the early 1990s and social media in the early 2000s. What print media cover (and leave out), how they cover, and how frequently and enduringly they cover all affect the discursive construction of nuclear power and potentially influence public opinion and public support for nuclear policy. This chapter, then, takes a close look at the coverage of nuclear power by two leading national newspapers, *The Yomiuri Shimbun* (*Yomiuri* hereafter) and *The Asahi Shimbun* (*Asahi* hereafter), from 1954 to 2000. *Yomiuri* (established in 1874) enjoys the largest circulation in Japan and is known to be conservative, whereas the second largest paper, *Asahi* (established in 1879), caters to more progressive, urban audience. As influential national papers, they are ideological apparatuses that have the power to construct truth (Foucault, 1984).

I chose to focus on editorials for two reasons. The first is feasibility; the number of articles published in the half century is enormous and is beyond the scope of this book. The second is the nature of editorials. Editorials are "opinion discourse par excellence" and are one of the most widely circulated opinion discourses in society (van Dijk, 1996). It is also important to remember that this opinion is not personal but institutional (van Dijk, 1996). Editorials express a newspaper's official stance on "the important political, economic and social news stories of the day, both domestic and international" (Yomiuri Shimbun Editorials, n.d.) and serve as the "face" of the newspaper (Editorial Board, *Asahi*, n.d.). Editorials thus offer salient insights into how these prominent papers choose to depict nuclear power. The editorials in Japanese newspapers are produced as a result of careful deliberations by an editorial board to represent the perspective of the newspaper as an organization and is thus published anonymously. As the top two national papers, the topics *Yomiuri* and *Asahi* cover matter.

DOI: 10.4324/9781003044222-3

Both papers published a substantial number of editorials on nuclear power. My initial archival research, using "genshiryoku" (nuclear power) and "nihon" (Japan) as keywords, yielded over 800 results or, on average, at least one editorial on nuclear power in Japan per month. This alone suggests that the dominant newspapers thought of nuclear power as one of the most important topics worthy of frequent analysis and response as an organization. I combed through the hundreds of articles and eliminated the ones that only marginally discussed Japan's nuclear power. For example, articles about international treaties and conferences on nuclear power, the nuclear power development in other countries, and Japan's future energy policy in which nuclear power is listed among others but not elaborated were eliminated. This process yielded 187 *Asahi* editorials and 271 *Yomiuri* editorials that substantively dealt with some aspects of nuclear power. The difference in the quantity between the two paper is noteworthy; out of the myriad of national topics, Japan's largest newspaper, *Yomiuri*, gives more attention to nuclear power than *Asahi*. This is no surprise; as mentioned in Chapter 2, *Yomiuri*'s then president, Matsutaro Shoriki, was a fervent supporter of nuclear power and became the first chair of the JAEC, the nuclear power regulatory body established in 1954.

The two papers are commonly understood to represent contrasting political proclivities. *Yomiuri* is conservative and sympathetic to the ruling, center-right LDP that has been dominant in post-WWII Japan, whereas *Asahi* is liberal-leaning and tends to support the center-left political agendas and activities. Based on this casual look of the difference between the two papers, it may be hypothesized that their editorials would be contrasting not only in the quantity but also in the content. Their perspectives, however, were not so incompatible. To discern the convergence and divergence between the two papers, I based my analysis around two questions: how did the papers depict the same topics, and what topics were addressed by only one paper and how were they depicted?

I used framing to examine the similarities and differences in the ways the two newspapers represented, dissected, and packaged the development of Japan's nuclear power program. According to a media communication scholar, Robert Entman (1993), framing involves two steps, *selection* of a portion of a perceived reality and giving it *salience*, and this framing occurs at four distinct locations within a communication process. Communicators frame their texts (what they say or write) by making conscious and unconscious decisions about what to say, and these decisions are guided by schemata that organize their belief systems. The texts contain frames in the form of "the presence of absence of certain keywords, stock phrases, stereotyped images, sources of information, and sentences that provide thematically reinforcing clusters of facts or judgements" (Entman, 1993, p. 52). The receivers of the texts bring their schemata in reading the texts and may or may not be influenced by the texts and the framing intention of the communicators. Finally, the culture is "the stock of commonly invoked frames" (Entman, 1993, p. 53). In all four locations, Entman argues, framing performs several functions: define problems, diagnose causes, make moral judgements, and suggest remedies. By performing these functions, news discourse builds shared knowledge, ideologies, and attitudes (van Dijk, 1995), thus participating in discursive formation of nuclear power (Foucault, 1972/2010).

For this study, my primary subjects of analysis are the texts – the newspaper editorials. The communicators – the editorial boards of *Asahi* and *Yomiuri* – and the culture are also considered secondarily. Although I do not have access to the intents of the communicators, their preferred frames can be discerned through the examinations of their texts and the rhetorical strategies that are deployed to perform the aforementioned functions. Such strategies may include overstatement, understatements, metaphors, hyperbole, euphemism, and mitigation (van Dijk, 1995) among others. The cultural frames are relevant to the extent that the ideas that are circulated in various public discourse spaces may well find their ways into texts as communicators utilize them to construct a frame in favor of their stories. I organized my analysis roughly by decade. This was primarily for convenience, and yet each decade interestingly presents a somewhat coherent character in the ways the editorials engaged nuclear power matters.

1954 to 1964: celebrating the dawn of Japan's nuclear era

The editorials concerning Japan's use of nuclear power began to appear in both *Yomiuri* and *Asahi* in early 1954 when the National Diet approved the 2.35 million yen budget for nuclear power development.[1] The first decade from the budget approval until the early 1960s was the dawn of Japan's nuclear era. The government moved swiftly to establish a political infrastructure to begin a nuclear power program. Accordingly, the editorials concentrated on topics around it: the formation of the administrative, industrial, and research bodies (e.g., the JAEC, the Japan Atomic Industrial Forum, the Japan Atomic Energy Research Institute), policies, and goals for nuclear power development. *Asahi* published 37 editorials, and *Yomiuri* published 36 editorials concerning Japan's nuclear power program. Although the editorials made disapproving observations such as delays, challenges, and inefficiencies as editorials generally do, the overall tone of both papers was hopeful and supportive of Japan entering the era of "peaceful use" of nuclear power as declared by the U.S. President Eisenhower in 1953. *Yomiuri* (1956, January 14) wrote that "becoming a nuclear power nation is an immediate necessity for the survival of Japan" – an argument that *Asahi* (1957, January 6) seemed to agree with when it wrote that "If nuclear power comes to make sense economically, humanity will no longer have to worry about the lack of electricity."[2]

At the same time, there were subtle differences between the two papers in how Japan was to proceed. The editorials from August 27, 1957 are a good example. On this day, both papers celebrated Japan's first research nuclear reactor reaching criticality at the Japan Atomic Research Institute

(JAERI). This was their top front-page news and was the topic of their editorials. *Asahi* called the occasion as "the first day nuclear power is no longer a theory but a reality for Japan," and *Yomiuri* dubbed it as "the dawn of nuclear Japan" and "an energy revolution." The two editorials were slightly different in tone. *Yomiuri*'s was entirely celebratory and was placed next to the largest headline, "the sun's fire is lit," and a "historical moment" message from Shoriki, whereas *Asahi* reminded readers that the whole technology was imported from the United States and cautioned that piggybacking on a foreign technology won't lead to an authentic "Japan's nuclear power." The *Asahi* editorial was on the second page next to a story about a political conflict over the establishment of a company that will import the first commercial reactor from the Great Britain, further directing the audience's attention to Japan's reliance on other nations.

This difference in tone continued through the first decade. On October 26, 1963, JAERI succeeded in generating electricity from an experimental reactor in Tokai-mura, Ibaragi, and both papers wrote an editorial about it. *Yomiuri* (1963, October 27) celebrated the occasion by remarking that, despite being ten years behind the western developed nations, Japan is the 11th country to generate electricity from nuclear power and that the small reactor, if it operates at the full capacity, produces electricity equivalent to the amount consumed in Tsuchiura, a city south of Tokai. It also forcasted that nuclear will soon replace oil as the cheap energy source. The only criticism the editorial made is the lack of creative cooperation among research facilities and corporations in nuclear power development projects. *Asahi* (1963, October 28) also praised the success in Tokai as "a promise of bright future" and "the first step of serious effort to advance nuclear power." The praise was short, and the rest of the editorial was cautious about embracing the milestone; it pointed out that the reactor's main parts and fuels are all imports from the United States, and, while it is important to learn from foreign technologies, the success at Tokai becomes meaningful only when Japan develops its own reactors. Aside from this variance in tone, both papers clearly supported nuclear power as a righteous technology for Japan and saw this first decade as the dawn of Japan's nuclear era.

1965 to 1978: need for nuclear power, need for public enlightenment

Support for breeder reactors

Moving into the second decade of nuclear power, the difference between the two newspapers in the frequency of the editorials about nuclear power is noteworthy. *Yomiuri* published 80, whereas *Asahi* published 38. This difference alone suggests that *Yomiuri* was far more interested than *Asahi* in communicating to their readers about the unfolding development of Japan's nuclear power program. In principle, however, both papers remained supportive of nuclear power as a necessary technology. As it was becoming apparent that uranium-235 needed for conventional nuclear energy generation is only 0.7% of natural uranium, the development of breeder reactors that convert abundant uranium-238 into plutonium-239 and use of it as a fuel were a topic of much interest for both papers. *Asahi* (1966, July 25) wrote that, despite many challenges, Japan absolutely needs breeder reactors and praised the nuclear industry for their plan to develop them. *Yomiuri* (1966, July 27) called breeder reactor a "dream nuclear reactor" that produces more fuels than it consumes and urged Japan to join other countries that are developing it. The editorial cited the industry's estimate that, in 20 years, 47% of Japan's electricity will come from nuclear power.

Antinuclear movements

Between 1966 and 1969, curiously *Yomiuri* was rather quiet about nuclear power in Japan, although they frequently wrote about nuclear development and nonproliferation overseas. In contrast, *Asahi* continued to print editorials on the subject matter. One topic that came up more than once was the difficulty of securing plant sites. The editorial from September 22, 1966, referenced the case of Chubu Electric Power Company's plan to build two plants in Mie Prefecture that was met by tough opposition from local fishermen. The editorial dismissed the fishermen's concerns by observing that, with today's advanced technology, it is unthinkable that a small problem at a plant will cause an explosion or disperse radioactive materials. *Asahi* suggested that the basic solution for the site acquisition problem is gaining cooperation and understanding from the local residents by conducting "unceasing public relation and enlightenment activities." *Asahi* repeats this message in their editorial on August 21, 1968 and urges the industry to acknowledge the importance of such activities.

After a few years of latency, *Yomiuri* wrote in their May 12, 1969 editorial that Japan is entering a true nuclear energy era, because, although Tokai Nuclear Power Plant is the sole plant at the moment, five reactors are under construction, including one in Tsuruga, two in Fukushima, and two in Mihama, and four more are slated for construction that fall. The same editorial commented that there are less safety-related opposition by locals because they now understand that plant construction abides by strict standards and inspections and thus would not harm their health. The editorial advises the authorities to ensure safety through constant monitoring.

Notwithstanding *Yomiuri*'s optimism about opposition, local resistance grew. Both papers began to address the importance of providing the public opportunities to ask questions and voice their opinions so as to gain their support for nuclear power. *Yomiuri* (1972, July 24) wrote that, while Japan has entered the era of practical application of nuclear power with four operating reactors and 14 reactors under construction, antinuclear movements are also getting strong because of safety and environmental concerns. Their editorials (1972, February 23 and July 24) argued that determination of nuclear facility safety should reflect citizens' voices and that public hearings should be held toward that end. *Asahi* (1972, June 2) underscored the importance of the nuclear industry following the three basic principles of peaceful use of nuclear power (democracy, independence, and public disclosure) and the particular importance of public disclosure in dealing with siting conflicts. This message was reinforced in the August 30, 1972 editorial that pointed out the growing opposition to plant construction due to the worsening environment. To deal with this problem, the editorial urged the authorities to legislate public hearings so the industry can share information about the safety of nuclear facilities and environmental protection and the citizens can voice their views.

The editorials' call for legislating public hearings in the target communities as a way to deter opposition and to gain public trust continues through late 1970s in both papers. *Asahi* (1978, October 4) and *Yomiuri* (1978, October 5) both acknowledged the establishment of the Nuclear Safety Commission (NSC) of Japan on October 4 as a right step toward ensuring nuclear power safety and expressed their hope for the NSC to be rigorous in enforcing safety standards and hold public hearings so as to gain public trust in nuclear power. *Yomiuri* (1978, November 4) wrote that "public acceptance" with regard to the nuclear industry has two meanings: general citizens' support for nuclear power and local residents' support for building plants in their towns. The majority of the Japanese citizens, the editorial observed, support nuclear energy but are opposed to having a plant in their town. The editorial suggested that the authorities institute a mechanism for all stakeholders to come together and engage in rational debates based on accurate information.

As shown, from the mid-1960s to the late 1970s, Japan's nuclear power industry entered the first decade of commercial energy production, and the construction of nuclear power plants gained momentum. As the need for more reactors grew, so did antinuclear movements. Opposition grew in communities that were targeted as new plant sites, delaying the construction goals of the nuclear power industry. The newspapers saw this struggle as a result of the lack of effort on the part of the industry and the authorities to be informative to the public and repeatedly called for instituting formal venues such as public hearings. Both papers also saw the uranium-235 shortage as a serious threat to Japan's nuclear power future and encouraged the industry to boost its effort to develop breeder reactors. Essentially, both papers believed in the necessity and goodness of nuclear power and saw nuclear safety as an administrative and poor public education problem. The solution is more transparency and dissemination of "accurate information" about nuclear power.

1979 to 1989: the era of accidents and parting ways

Starting with the TMI accident, this was the decade of nuclear disasters. Not surprisingly, in general the frequency of media coverage of nuclear power dramatically increased after accidents, and tones changed over time (Kristiansen, 2016; Gamson & Modigliani, 1989), including newspaper editorials. Sixty-five and 80 editorials were published by *Asahi* and *Yomiuri* respectively regarding Japan's nuclear power, and unsurprisingly the accidents appeared in many of the editorials. In addition, as opposition grew in the communities sited for nuclear power facilities, both papers devoted considerable attention to that topic.

The TMI accident and enduring support for nuclear power

The TMI accident on March 26, 1979 came as a serious blow to Japan's nuclear culture. At the time of the accident, Japan owned 18 reactors, and all but one of them used the TMI-type light-water reactors built with the U.S. technology. *Yomiuri* and *Asahi* published similar editorials, evaluating the impacts of the accident on Japan. Both papers urged the nuclear authorities to develop comprehensive safety and evacuation plans and suggested the need for Japan to advance its own nuclear technology to enable swift responses to accidents and breakdowns. While safety was a preoccupation of both papers following the accident, the editorials a year after the accident show divergent foci of the two papers. *Asahi* (1980, March 29) brought up the serial occurrences of accidents and troubles at several nuclear power plants, including grave ones at the Ōi No. 1 reactor and the Takahama No. 2 reactor, and resultant low operation rate (less than 50%). The editorial linked the frequent troubles to the rushed constructions of reactors and the nuclear industry's tendency to pursue newer and larger reactors without allowing new technology to take roots. It also criticized the industry for failing to change the claim of "nuclear is cheap" when it is in fact expensive if the enormous costs of accidents like TMI are taken into account.

The *Yomiuri* editorial from the same period (1980, March 28) sharply contrasts *Asahi*'s. After cursorily recognizing the changes the nuclear authorities are making in response to TMI, the editorial dedicated most of its space commenting on the sluggish expansion of nuclear power sites. In the last ten years, the number of the prefectures that host nuclear power plants increased only by three (from seven to ten) although the number of reactors tripled. The average plant construction time (from the time a formal request for construction is made to a town to the first day of operation) had nearly doubled to above 15 years. To remedy these site troubles, *Yomiuri* encouraged the nuclear authorities to make it easy for local residents to accept a nuclear power plant by legislating a public hearing system and implementing strategies to connect nuclear power to local welfare. It questioned, for example, whether massive hot wastewater from the plant could be used to heat the community. While *Asahi* evaluated the state of Japan's nuclear power to be premature and began to raise questions about its economic feasibility, *Yomiuri* saw the state to be not moving fast enough. At this time, however, neither paper questioned the feasibility of nuclear power from safety and environmental standpoints.

To be sure, both papers continued to write about the need for constant attention to safety. For example, in late July of 1984, the residents who had sued the government to stop the Fukushima Daini Nuclear Plant for safety reasons lost. Six years earlier, Ikata residents brought a similar law-suit and lost. Both *Asahi* (1984, July 24) and *Yomiuri* (1984, July 24) similarly observed that the legal determination that the safety test of the nuclear facility was legally adequate does not mean that safety is guaranteed and that the local residents' concern for safety is understandable after seeing the TMI accident. *Yomiuri* also noted that there were 27 accidents and break-downs at nuclear facilities in 1983 alone, a number that the paper found significant. *Asahi* questioned the adequacy of limiting the safety issue to facilities and excluding human errors when the TMI accident proved them to be fatal. Both papers concluded that, as Japan's reliance on nuclear power will likely to increase in the future, safety must be the top priority. While showing concerns for safety, both papers principally endorsed the continuing growth of nuclear power, and their editorials did not go beyond calling for strengthening safety measures.

In December 1985, just four months before the Chernobyl accident, each paper wrote to acknowledge the 30th year anniversary of nuclear power by recognizing its growth. In 30 years, nuclear power grew to contribute a quarter of Japan's total electricity and became cheaper than thermal and hydro power. *Yomiuri* (1985, December 5) noted that Japan currently has 31 reactors, and most were built with Japan's own technology. The challenge now, the paper observed, is accomplishing the use of plutonium, a byproduct of power generation at light-water reactors. Similarly, *Asahi* (1985, December 6) called for enhanced effort into securing fuels. It pointed out that the fuel cycle – recycling of spent fuel by reprocessing it – has not been achieved due to the difficulty of finding sites in the face of strong anti-nuclear movements in the 1970s. This changed in April 1985 when Aomori Prefecture and Rokkasho Village municipality in Aomori agreed to host facilities for uranium enrichment, reprocessing, and waste storage. *Asahi* applauded this development and argued that the authorities must proceed carefully and make efforts to gain support of the residents who oppose the facilities.

Citizens' role in nuclear power site decisions

What role should citizens play in making decisions about nuclear power plant sites? This has been a lingering question from the very beginning of

Japan's nuclear power development. The Basic Atomic Energy Law, established in 1955, set out three basic principles of peaceful use of nuclear power: democracy, independence, and public disclosure. Citizens' participation is implied in democracy and public disclosure principles. As shown in the preceding section, newspapers began to emphasize the importance of reflecting citizens' voices in nuclear power development in early 1970s. *Asahi* and *Yomiuri* both urged the nuclear industry and authorities to create opportunities for the public to learn about the safety and environmental impacts of nuclear facilities and to express their opinions and concerns. The papers repeatedly advocated for using formal public hearings for this purpose. Similarly, the papers favorably responded to the science symposium held to bring experts and academics together to discuss lessons from the TMI accident and called for more symposiums that include operators (*Asahi*, 1979, November 28) and non-experts (*Yomiuri*, 1979, November 29).

The difference between the two papers became notable with regard to referendums as a tool for public participation in the construction of nuclear power facilities. The first town where a referendum was proposed over the construction of a nuclear power plant was Kubokawa, Kochi Prefecture. After some years of conflict, the incumbent mayor, who first opposed the plant construction, was recalled because he switched sides. Then he was reelected upon promising that he would institute a referendum to resolve the matter. Both papers responded to this political development. When the mayor was recalled through an election in March 1981, *Asahi* (1981, March 10) wrote that there is much to learn from the election result. One lesson is that, unlike previous site fights, Kubokawa's antinuclear movement was led by the alliance of the local fishermen, women, and young dairy farmers who mobilized residents through hundreds of small meetings, whereas the pronuclear side made it a political spectacle and brought the LDP leaders as preachers. This politicization put off the local residents. The second and the third lessons are the ineffectiveness of the attitude of the nuclear industry and authorities that the site fights can be resolved through money and power. The recall result, *Asahi* concluded, was an important democracy lesson that local residents' views must be respected. In contrast, *Yomiuri* (1981, March 10) wrote that, although the mayor recall election in Kubokawa received a majority, it is not necessarily based on the public's sufficient understanding of nuclear power. To remedy this knowledge deficit, the government must promote public awareness from the very beginning, legislate public hearings, and hold nuclear symposiums where the experts can debate rather than speaking to residents at local fights and confuse them as it happened at Kubokawa.

The contrasting assessments of referendums continued the following year when a bill to hold a referendum to decide whether to host a nuclear power plant was submitted to the Kubokawa Town Council. *Asahi* (1982, June 30) called this referendum a valuable experiment for how to best reflect local residents' perspectives and observed that the success of the referendum rests on whether the mayor can keep his word and whether the whole process can be free from unfair financial and political influences. *Yomiuri* (1982, July 1), on the other hand, was doubtful of a referendum as an appropriate tool for settling a nuclear site fight. The editorial argued that a referendum can threaten democracy by privileging the wishes of the local town over the wishes of the larger public and by undermining the representative democracy (the town council). This contrast between *Asahi* and *Yomiuri* would surface again in the mid-1990s.

Apart from this difference over nuclear referendums, the two papers remained supportive of public hearings as a key approach to reflect public voices in nuclear power development in the following years. A public hearing regarding the construction of the second reactor at the Shimane Nuclear Power Plant became the first public hearing in the country in which the opposition also participated. Thirteen public hearings on nuclear power were previously held elsewhere, but the opposition refused to participate in them. Both papers praised this event as a progress. *Yomiuri* (1983, May 17) wrote that, although there were flaws (e.g., simple presentations without time for dialogue), the participation by both sides of the debate was a first step toward meaningful public hearings. *Asahi* (1983, May 16), while also pointing out flaws, called the hearing "epochal."

The Chernobyl accident

It was in the midst of this slow move toward public engagement when the news of Chernobyl accident inundated the media across the globe on April 26, 1986. Both *Asahi* and *Yomiuri* published multiple editorials in the subsequent months, trying to make sense of not only what transpired at Chernobyl but also its implications for Japan's nuclear power. Both papers recognized the accident as a result of human errors and noted the lack of experience and knowledge as the main reason that led to the fatal accident (e.g., *Asahi*, 1986, June 20 and August 17; *Yomiuri*, 1986, June 6 and August 21). The editorials called for the creation of international safety standards and the need for advancing knowledge and skills of technicians. After these commonalties, however, the two papers' views diverged. The contrast is clear in the editorials both papers published on August 31, 1986. *Yomiuri* emphasized the need to overcome the tragedy and regain public trust in nuclear power by strengthening commitment to safety. In contrast, *Asahi* referenced the paper's most recent public opinion poll that showed more opposition to nuclear power (41%) than support (34%). *Asahi* conducted a similar poll six times previously, and this was the first time the opposition was stronger than the support. *Asahi* attributed this change to Chernobyl and suspected that the public has begun to question the very premise of nuclear power safety.

The year following the accident, *Yomiuri* continued to write in support of the nuclear program. The paper suggested that the divided international responses to Chernobyl are an opportunity for Japan to show leadership in nuclear power technology advancement (1986, November 6). On the one-year anniversary of Chernobyl, *Yomiuri* wrote that nuclear power provides enormous electricity cheaply and fights global warming and found a French nuclear authority's words – reactors can easily last 50 years by updating parts – very encouraging (1987, April 20). A month later, they commented that Japan's reactors are highly reliable and safe and technicians are highly trained to operate the machines (1987, May 30). While *Asahi*'s editorials remained rather latent about nuclear power in the year following the disaster, both papers' responses to the year-end nuclear power white paper published by the JAEC were similar. They recognized that Japan now produces 28% of its electricity from nuclear power, making it the fourth largest nuclear power nation following the United States, the Soviet Union, and France and called for advancing nuclear safety to regain public trust (*Asahi*, 1987, December 7 and *Yomiuri*, 1987, December 2). This was, however, the last supportive editorial *Asahi* published until the mid-1990s.

To be sure, both papers published hundreds of articles about Chernobyl, but the majority of them were about the accident and its impacts on the local citizens around Chernobyl and did not articulate the implications of the accident for Japan's nuclear power program (Yamakoshi, 2015). That is, the accident was represented as "the fire across the other side of the river" or someone else's problem. When radiation from Chernobyl reached Japan in May 1986, the problem suddenly hit close to home. However, the Japanese government issued a statement that there is no risk to Japan, and both papers briefly covered it (Yamakoshi, 2015).

After curiously remaining dormant during 1987, *Asahi* published ten editorials on nuclear power from 1988 to 1989. Its editorial from April 26, 1988, the two-year anniversary of Chernobyl, read: "we accepted the necessity of nuclear power as a replacement to oil on the conditions that it is limited to peaceful purposes, is safe and economical, is managed responsibly, and is accepted by local communities" and called out the failure of satisfying these conditions. This failure is brought up in subsequent editorials: for example, the nuclear authorities and industry's lack of considering residents' fear, distrust, and anger (1988, July 6; 1988, August 14); public distrust of nuclear safety and nuclear technology (1988, September 28); the problem of decommissioning old reactors (1988, November 1); growing antinuclear movement (1989, April 30); and the costs of decommission and nuclear waste management (1989, November 18). *Asahi*'s active and antagonistic coverage of nuclear power reflected the continuing challenges of Chernobyl cleanup and health effects as well as the increase in the publication of antinuclear books such as Takashi Hirose's *Kiken na hanashi* (the dangerous story) that delineated the magnitude of the Chernobyl accident and raised questions about the safety of Japan's nuclear reactors and the nature of nuclear power industry (Yamakoshi, 2015).

These considerably differed from a dozen of editorials published by *Yomiuri* in the same two years. *Yomiuri* maintained the characterization of nuclear power as an indispensable and best energy source for Japan, citing finite fossil fuels and impracticality of new energy sources (solar, geothermal, and wind) (1988, October 14 and December 4), the highest safety standards of Japan's nuclear program (1988, October 14 and December 4; 1989, March 14), nuclear power as the answer to the global environmental crisis (1988, October 14, November 8, and December 4; 1989, October, 27 and November 18), and affordability (1988, June 11). *Yomiuri* also wrote about the fear and distrust expressed by the public. In contrast to *Asahi*, however, *Yomiuri* saw them as a manifestation of the lack of correct understanding of nuclear power (1988, February 15) and a response instigated by the antinuclear propaganda (1988, June 11).

A review of the editorials from the late 1970s through the 1980s showed an interesting shift from a general alignment to divergence between *Asahi* and *Yomiuri*, indicating different responses to the two foreign nuclear accidents. The TMI accident prompted both papers to underscore tighter safety measures and the need to engage the public through public hearings. *Asahi* also welcomed public engagement through referendums as an exercise of democracy, a view *Yomiuri* did not share. With some difference, both papers in principle continued to support nuclear power as a necessary technology whose challenge lies not so much in its very design but its lack of public engagement and education. This alignment was no longer the case after Chernobyl. While *Yomiuri* continued to write in support of the nuclear program, *Asahi*, for the first time, began to question the nuclear program itself. The last editorials both papers published on November 18, 1989, in response to the annual nuclear white paper illustrate the different paths the papers began to take. *Yomiuri*'s argument was two-fold: nuclear power is indispensable in fighting global warming, and public understanding and cooperation must be won. *Asahi* questioned the premise of the inevitability of nuclear as *the* replacement to fossil fuels; it pointed out that nuclear is dubbed reliable and affordable, but the price does not include the costs of decommissioning the old reactors and maintaining the spent fuels. This divergence continued into the 1990s.

1990 to 1999: divergent paths

In the last decade of the 20th century, *Asahi* published 47 editorials on Japan's nuclear power. They wrote about a variety of topics: safety concerns, nuclear power's role in Japan's overall energy policy, the accidents that occurred during the decades, the legacy of the Chernobyl accident, the lack of transparency in nuclear administration, the nuclear fuel cycle and construction of a fast breeder reactor, democracy in nuclear power decisions, and nuclear waste. The tone of the editorials remained consistently critical across the topics; they questioned the viability of nuclear power and expressed concerns about Japan's increasing reliance on it. Their editorial from December 2, 1992, exemplifies *Asahi*'s post-Chernobyl position on nuclear power: "Japan received the baptism of nuclear bombs and yet pursued nuclear power and intends to continue this reliance. . . . The nuclear power era began a half century ago. Does this era signal human progress, or is it a colossal human foolishness? The jury is still out."

Yomiuri published 75 editorials during the decade concerning Japan's nuclear power. Some of the topics were the same as *Asahi*: safety, nuclear fuel and the fast breeder reactor construction, accidents, referendums, and nuclear waste. *Yomiuri* also frequently referenced the role of nuclear power in the fight against global warming, a topic that *Asahi* rarely took up. All in all, *Yomiuri* maintained their unwavering support for nuclear power, characterizing it as the energy source indispensable for resource-poor Japan and for reducing carbon dioxide for the Earth. The divergent paths that the two papers began to take after Chernobyl continued through this decade.

Making sense of accidents: faith versus doubts in nuclear technology

The ways the two papers framed accidents offer a glimpse of the contrast between them. The first accident of the decade occurred on February 9, 1991, at the Mihama Nuclear Power Plant in Mihama-cho, Fukui Prefecture. A metal tube in the stream generator of Unit 2 broke, and a small amount of radiation was released. *Asahi* declared that this was a serious accident that threatened the safety of Mihama's pressurized water reactor (PWR) (1991, February 11) and analogized the incident to an annual health exam; you got gravely ill even if you had just seen a doctor for a checkup, and, when you complained to the doctor why they missed it, they would tell you that the problem you have was not part of the assessment (1991, March 13). *Asahi* saw this as a "fundamental flaw" of nuclear power safety management. *Yomiuri* criticized the plant for its poor safety inspections but emphasized that the management of the accident was a textbook case (1991, February 11) and argued that the authorities must tighten nuclear safety and gain public trust because nuclear power is a vital energy source for Japan (1991, March 13). It reminds the readers that the only reason Japan can stay calm despite the Gulf War is because we have nuclear power (1991, March 13).

A number of accidents occurred after the Mihama accident, the most severe of which occurred on September 30, 1999, at the JCO uranium-processing plant owned by JCO in Tokaimura, Ibaragi Prefecture. The errors the technicians made while preparing uranyl nitrate caused an uncontrolled nuclear chain reaction for the next twenty hours, exposing three technicians to heavy radiation doses. Two of them died. Both papers published multiple editorials following the accidents. *Yomiuri* concluded that the accident was caused by the unlawful activities and dismal safety management of the processing plant (1999, October 5 and November 4) and concluded that the nuclear industry must learn lessons and avoid future accidents and activities that shake the public confidence in nuclear power (1999, October 1 and 5). "The public and private sectors," wrote *Yomiuri*, "must work together to rebuild the safety management system in order to regain the public confidence in nuclear power. Energy resource poor Japan needs nuclear power" (1999, December 25). *Asahi* wrote about the citizens' fear (1999, October 1) and the fear expressed by other Asian nations about the fundamental flaws of nuclear power that even an advanced nation such as Japan is unable to avoid (1999, October 4). In sharp contrast to *Yomiuri*'s December editorial, *Asahi* remarked,

> On the ground that nuclear power is indispensable to resource poor Japan, the government has actively promoted construction of nuclear power reactors and has pursued a nuclear fuel cycle to use the amassed plutonium resulting from the spent fuels. This was the only option they entertained. But we are at a crossroad. We must consider the possibility of reducing nuclear power and terminating the plutonium fuel cycle program.
>
> (1999, December 23)

Return to referendum

Another topic that clearly demonstrates the difference between two papers is referendums. In 1996, Maki-machi, Niigata Prefecture, implemented a referendum regarding the construction of a nuclear power plant. This was the first referendum of this kind in the country, and the opposition won by a considerable majority. *Asahi* (1996, August 5) wrote:

> we hope more municipalities adopt referendums following Maki. . . . We don't know what implications Maki's result may have on other nuclear power plant sites. But one thing is clear; this experiment of direct democracy exposes the inefficacy of Japan's representative democracy and demands deep reflections on the political system.

Yomiuri also published an editorial on the referendum on the same day: "The construction of nuclear power plants is a matter of national energy policy. Local referendums should not determine a national policy. The government must make an effort to acquire local understanding about nuclear safety." While *Asahi* saw referendums as a brilliant method of directly reflecting citizens' wills (1996, January 26), *Yomiuri* viewed it as a problematic tool for nuclear power matters because "it is a tool favored by the antinuclear movement" and because "national policy issues should not be decided by local residents just because they don't want the plant in their communities" (1996, July 14). The different stances with regard to referendums reveal the two papers' contrasting beliefs about what power the citizens should have about informing the nuclear power program and public health concerns that emanate from the program.

Conclusion

After the Fukushima Daiichi accident, *Asahi* ran an investigative report series called *Purometeusu no wana* (The trap of Prometheus) from October 2011 to March 2016. The series began with a critical examination of the accident itself – the cause, context, and aftermath – and expanded its scope to address a larger question – does Japan really need nuclear power? The series won multiple awards and was later published as a series of books. The title is revealing; nuclear power was likened to the Greek god and trickster who brings fire to humanity. Fire represents technology that represents human progress and the demise of unintended consequences brought on by the technology. This series unambiguously represented the antinuclear power stance of *Asahi* in post-Fukushima Japan. As this chapter demonstrated, this was not always the viewpoint *Asahi* held. While *Asahi* and *Yomiuri* are often understood as news outlets with contrasting political leanings, their views of nuclear power were more similar to each other than not for the first three decades of Japan's nuclear power history. The tone of *Asahi* was generally more cautious and critical than *Yomiuri*, but in principle they both endorsed nuclear power as a technology necessary for natural resource-poor Japan. This positive depiction of nuclear power is not unique to Japan but is a global one (Kristiansen, 2016). In his study of newspaper coverage of the Fukushima Daiichi accident, Yamada (2016) observes that *Asahi* and *Mainichi* took antinuclear stances while *Yomiuri* was pronuclear and that this difference was already somewhat present in the 1960s. My analysis of the editorials confirms that the differences in tone were indeed present, but not visibly and consistently so until around 1988.

It was only after the Chernobyl accident that the positions of *Asahi* and *Yomiuri* began to clearly diverge, and it was the general trend in the 1990s. *Yomiuri* was critical of frequent accidents, inefficiencies, and poor safety management, but they continued to argue for the necessity of nuclear power as the best energy source for Japan and for fighting global warming.[3] *Asahi*'s post-Chernobyl editorials included similar topics, but they began to challenge the assumption that nuclear power was necessary and is a righteous technology. They became skeptical of the idea that safety can be guaranteed by better technology and controlling human errors. They also questioned the cost-effectiveness of nuclear power – a frame that was also absent in the U.S. nuclear power discourse before Chernobyl (Gamson & Modigliani, 1989). *Asahi* of course was not alone on this path. For example, a number of antinuclear power television documentaries were aired in the 1990s, exposing site fights and manipulative and forceful strategies used by electric companies to acquire lands and consent of town leaders and residents (Karasudani, 2014) and presenting local communities' perspectives on the hardships and fears about hosting nuclear facilities (Seo, 2018).

By the time *Asahi* (and other liberal papers such as *Mainichi Shimbun* and *Tokyo Shimbun*) began to write regularly about the fundamental flaws of nuclear power, including the inevitability of human errors, casualties that accompany severe accidents, vulnerability to earthquakes, unsolved nuclear waste problems, and more, nuclear power had deep and extensive roots in Japan's political, economic, and social landscapes. It was providing over a quarter of Japan's electricity, and, as *Yomiuri* repeatedly hailed, its story as the most promising clean and green energy for Japan was dominant. As with pronuclear discourse elsewhere (Windisch, 2008), nuclear power was hailed as the most promising solution to global warming. Would Japan have traveled a different energy path if liberal media were critical of nuclear power from the beginning? Studying the nuclear news published by *Asahi* and *Yomiuri* between 1945 and 1965, Yamamoto (2014) argued that newspapers helped to shape nuclear power dreams. For the majority of Japan's nuclear power history, newspaper editorials helped to shape the dreams, framing nuclear power not as a cousin of nuclear bombs dropped on Hiroshima and Nagasaki but as a virtuous technology of peace and prosperity.

According to *Asahi* journalist Yoichi Jomaru, *Asahi* historically took a "yes, but" stance regarding nuclear power; it approved nuclear power in principle but would criticize its safety and other problems (Kobayashi, Jomaru, Fackler, Endo, & Tanihara, 2012). Jomaru sees this stance as somewhat a reflection of the political struggle within the paper but largely as a mirror of the public opinions that simultaneously indicated fear and support of nuclear power. While this insight is useful, it must be noted that mass media is, as van Dijk (1995) argues, a powerful shaper of public opinions. The representational power of discourse (Foucault, 1972; Hall, 1997; Deetz, 1992) suggests that the discourse advanced by the powerful media shapes how nuclear power may be meaningfully and legitimately talked about while ruling out other ways of talking and other meanings. Public attitudes can be influenced and even controlled through dominant news media discourse, topics, meanings, styles, and rhetoric especially in the absence of alternative discourse. If we agree that the media is a powerful framer of the social world, it is not difficult to imagine that the national pronuclear discourse advanced by the influential newspapers over the decades has shaped the ways the Japanese public understood nuclear power.

Notes

1 Around the same time, both *Yomiuri* and *Asahi* began media campaigns to introduce the "peaceful use" of nuclear power. Both papers published a series that painted a remarkable, automated, convenient future nuclear power can bring to humanity (Yamamoto, 2014).

2 All quotes are the author's translations.
3 These are general tendencies shown in the editorials. When these newspapers write about nuclear power in different sections, their coverage and framing could be different. For example, Baku Nishio, director of the CNIC, observed that he can find sound analysis of nuclear accidents in the society section of *Yomiuri* whereas its science section frames nuclear power favorably (Personal communication, June 13, 2013).

References

Anderson, B. (1991). *Imagined communities* (Rev. ed.). London: Verso.

Deetz, S. A. (1992). *Democracy in an age of corporate colonization*. Albany, NY: State University of New York Press.

Editorial Board, Asahi. (n.d.). Retrieved from www.asahi.com/shimbun/honsya/e/e-edit.html

Entman, R. (1993). Framing: Toward clarification of a fractured paradigm. *Journal of Communication*, *43*(4), 51–58. https://doi.org/10.1111/j.1460-2466.1993.tb01304.x

Foucault, M. (1972/2010). *The archaeology of knowledge and the discourse on language* (A. M. Sheridan Smith, Trans.). New York: Vintage Books.

Foucault, M. (1984). *The Foucault reader* (P. Rabinow, Ed.). New York: Pantheon.

Gamson, W. A., & Modigliani, A. (1989). Media discourse and public opinion on nuclear power: A constructionist approach. *American Journal of Sociology*, *95*(1), 1–37. https://doi.org/10.1086/229213.

Hall, S. (1997). The work of representation. In S. Hall (Ed.), *Representation: Cultural representations and signifying practices* (pp. 15–64). Thousand Oaks, CA: Sage Publications, Inc.

Karasudani, M. (2014). Genshiryokuto terebi dokyumentari [Nuclear policies and TV documentaries]. *Mass Communication Kenkyu*, *84*, 29–51. https://doi.org/10.24460/mscom.84.0_29

Kobayashi, N., Jomaru, Y., Fackler, M., Endo, K., & Tanihara, K. (2012). Genpatsu hōdō no kenshō [Examination of nuclear power news coverage]. *Mass Communication Kenkyu*, *81*, 41–64. https://doi.org/10.24460/mscom.81.0_41

Kristiansen, S. (2016). Characteristics of the mass media's coverage of nuclear energy and its risk: A literature review. *Sociology Compass*, *11*, 1–10. https://doi.org/10.1111/soc4.12490

Seo, H. (2018). Rokaru kyoku ha genshiryoku mondai wo ikani tsutaetaka [How local stations reported the nuclear power problem]. *Mass Communication Kenkyu*, *93*, 97–115. Retrieved from www.jstage.jst.go.jp/article/mscom/93/0/93_97/_article/-char/ja

van Dijk, T. A. (1995). Discourse analysis as ideology analysis. In C. Schäffner & A. Wenden (Eds.), *Language and pace* (pp. 17–33). Aldershot: Dartmouth.

van Dijk, T. A. (1996). *Opinions and ideologies in editorials*. The 4th International Symposium of Critical Discourse Analysis, Language, Social Life and Critical Thought, Athens. Retrieved from www.discursos.org/unpublished%20articles/Opinions%20and%20ideologies%20in%20editorials.htm

Windisch, U. (2008). Daily political communication and argumentation in direct democracy: Advocates and opponents of nuclear energy. *Discourse Society*, *19*(1), 85–98. https://doi.org/10.1177/0957926507083690

Yamada, K. (2016). Fukushima Daiichi genshiryoku jiko wo meguru jyohogen to shimbun media no rendosei ni tsuite. *Waseda Daigaku Seiji keizai gakubu kyoyo-shogaku kenkyu kai, 141*, 111–139. Retrieved from http://fukushimastudy.org/study/

Yamakoshi, S. (2015). Cherunobuiri Genpatsu Jiko ni kansuru media gensetsu no bunkai. *Media Komyunikeishon, 65*, 17–27. Retrieved from www.mediacom.keio.ac.jp/wp/wp-content/uploads/2015/04/2015Yamakoshi.pdf

Yamamoto, A. (2014). "Genshiryoku no yume" to shimbun: 1945–1965 nen ni okeru Asahi Shimbun Yomiuri Shimbun no genshiryoku hodo ni kansuru ikkosatu [The dreams of nuclear power and newspaper: A study of nuclear power reporting by *The Asahi Shimbun* and *The Yomiuri Shimbun* in 1945–1965]. *Masu Komyunikeishon Kenkyuu, 84*, 9–27. Retrieved from www.jstage.jst.go.jp/article/mscom/84/0/84_KJ00009002044/_article/-char/ja/

Yomiuri Shimbun Editorials. (n.d.). *Corporate information*. Retrieved from https://info.yomiuri.co.jp/english/editorials.html

4 Pronuclear power discourse
Safe, indispensable, and green[1]

The TMI accident in 1979 and the Chernobyl accident in 1986 showed the Japanese public that nuclear power is dangerous. As we learned in Chapter 2, 3.6 million people signed the petition to discontinue the nuclear power program after Chernobyl. There were also a series of smaller nuclear accidents in Japan that further damaged the image of nuclear power. The nuclear-industrial complex needed to work on symbolic damage controls to restore the public's faith in nuclear power. This chapter examines the discursive tools that the government, the utilities, and the media used to promote positive images of nuclear power from the late-1980s until the Fukushima nuclear accident in 2011. In Chapter 3, we saw that dominant national newspaper editorials framed nuclear power as important and necessary for the majority of the nuclear power history. What other agents were at work to build and rebuild the public's trust in nuclear power and what discursive strategies did they deploy? This is the question the chapter seeks to answer. More specifically, I discuss prominent articulations through which nuclear power became hegemonic in Japan.

Articulation and hegemonic discourse

Articulation is a communicative act that connects two elements to establish a socially meaningful unity. Critical cultural theorists define articulation as a suture (Hall, 1985), a temporary unity (DeLuca, 1999) or a moment of closure (Slack, 1996). What these definitions have in common is contingency and historical particularity. The unity between the elements is contingent and has "no necessary correspondence" (Hall, 1985, p. 94). That is, the unity is not natural but is politically and ideologically produced in a particular historical moment and may change when called for by a changing historical context that contains specific political, ideological, and social forces.

Articulation is simultaneously a theory, politics, and a method of inquiry (Slack, 2006). As a theory, it is "a way of characterizing social formation without falling into the trap of reductionism and essentialism" (Slack, 1996,

DOI: 10.4324/9781003044222-4

p. 112). In other words, articulation theory allows claims to be made about identities, practices, and institutions while resisting the claims to assume monopoly and permanency. As politics, articulation empowers or disempowers certain ways of imagining and acting (Hall, 1985; Slack, 2006). That is, articulations are not simply ways of representing the social world, but they guide the ways we respond to the social world. Articulations and disarticulations, therefore, have social consequences. Finally, as a method, articulation suggests that we pay attention to the ways in which non-corresponding elements come together to create realities (Slack, 2006).

Articulation as theory and method is especially instructive in studying hegemonic discourse because hegemony relies on successful and enduring articulations. Williams (1977) defines hegemony as lived dominance and even culture. The relations of domination and subordination deeply saturate culture, including economic, political, and social activities, identities and relationships, so much so that the relations are seen as common sense or *the* reality for most members of a culture. Hegemony, thus, is power through consent (Artz & Murphy, 2000) where people are willing participants in maintaining the reality that benefits the dominant. The consent occurs because people self-identify with "the hegemonic forms: a specific and internalized 'socialization' which is expected to be positive, but which, if that is not possible, will rest on a (resigned) recognition of the inevitable and the necessary" (Williams, 1977, p. 118). I regard Japan's nuclear power as a form of hegemony, for it has deeply penetrated Japanese society not only economically and politically but also culturally to the extent that its necessity and virtue are largely unquestioned by the majority of the citizens, at least up until the Fukushima nuclear disaster.

Hegemony depends on articulations but requires more than articulations. Laclau and Mouffe (1985/2001) insist that hegemonic articulations occur only when antagonistic forces and frontier effects are present. Antagonisms are limits, differences, or conflicts within hegemony that must be articulated as such if subversion of the hegemony is to occur (DeLuca, 1999). Without antagonistic articulatory practices, there is no need for hegemony. Likewise, hegemony does not exist without floating elements that hegemonic articulations must work to stabilize.

Since hegemony is not permanent, it must be constantly defended, revised, and renewed through articulations, and a number of discursive strategies may be mobilized. Hegemonic articulations may link positive meanings to the existing structure (Art & Murphy, 2000) while suppressing or explaining away articulations that challenge them (Williams, 1977). Domestication, or the use of friendly and ordinary expressions to present disagreeable matters such as nuclear weapons and wars (Schiappa, 1989), may be another rhetorical strategy to achieve discursive control. Additionally, hegemonic articulations may utilize discursive closure strategies (Deetz, 1992) such as disqualification or exclusion of speakers, naturalization of meanings, neutralization of value-laden activities, topical avoidance, subjectification of experience (i.e., "it is a subjective matter"), and meaning denial to achieve hegemony.

Japan's pronuclear discourse summoned many of these discursive strategies. In the remainder of the chapter, I discuss four articulations that served as building blocks of Japan's pronuclear discourse by suturing nuclear power to righteousness and necessity. While hegemony requires consent of the masses, my analysis largely focuses on the discursive work by the dominant, nuclear-industrial complex that includes the nuclear power industry, the government, and the mainstream media. I do so because it is the nuclear-industrial complex that has extensive power to control discourse by producing and disseminating nuclear power knowledge, whereas ordinary Japanese have participated in pronuclear discourse largely as consumers and messengers of the knowledge.

In discussing the four articulations, I drew from a variety of sources including newspaper articles, nuclear power industry websites, government websites, television commercials, popular and scholarly books, and personal communication. Pronuclear discourse has been both shaped and challenged by many discursive texts and nondiscursive means. As such, the sources that informed my analysis are not exhaustive but are meant to illustrate the articulations.

Pronuclear articulations

Atoms for peace: Genshiryoku, not Kaku

I begin my analysis with an often overlooked but critical issue of naming. The Japanese language has two words – *kaku* and *genshi* – both of which mean "nuclear." *Kaku* originally meant "core" or "center" and was used to refer to the central part of something such as an issue, a group, or a social movement. However, in the context of WWII, *kaku* became "atom" and was employed as shorthand for nuclear armament, and this meaning has become the dominant use of the word. Thus, *kaku*, in practice, is synonymous to *"kaku bakudan"* (nuclear bomb) and *"kaku danto"* (nuclear warhead) and evokes the nuclear bombing of Hiroshima and Nagasaki. This association between *kaku* and nuclear weapons was evoked again when the U.S. nuclear testing exposed the crews of *Daigo Fukuryu Maru* to acute radiation in 1954 (see Chapter 2). To confront this antagonism, the pronuclear bloc articulated nuclear dualism – separation between atoms for peace and atoms for destruction (Doyle, 2011) by adopting a different name, *genshiryoku* – *genshi* ("nucleus" or "atom") and *ryoku* (power) – for nuclear energy. According to Yasuhiko Yoshida (2007), a former director of the International Atomic Energy Association, the use of two separate words was a purposeful effort by *genshiryoku mura* ("the Nuclear Village"),[2] or the nuclear-industrial complex, to ensure that the citizens do not link nuclear power to atomic bombs but instead see it as something agreeable.

In this suturing of *genshiryoku* to Japan's better future, mass media played a critical role. First, newspapers served as a powerful medium. Starting in 1954, Japan's largest newspaper, *The Yomiuri Shimbun*, began an aggressive campaign to promote peaceful use of nuclear power by writing about the vast possibilities of nuclear power for peaceful use, the increasing reliance on nuclear power in Europe and the United States, and the need to catch up with these countries (Arima, 2008). It was no accident that *The Yomiuri Shimbun* became an eager campaigner for nuclear power; it was owned by Matsutaro Shoriki, a lower house member who, along with Nakasone, came to be called the founding fathers of Japan's nuclear power. But other papers, too, chimed in. For example, *Tokyo Shimbun*, a metropolitan newspaper widely circulated in and around Tokyo, published a story on December 31, 1955 that pointed out the finite nature of natural resources and lauded *genshiryoku* as a peaceful resource indispensable to human survival (Koide, 2012).

Hiroaki Koide, a nuclear power and radiation expert, bears witness to the success of the media articulations of atoms for peace:

> When U.S. occupation of Japan ended as a result of the San Francisco Peace Treaty, the media began to report how horrific nuclear bombs were. But, at the same time, they advertised enormous potentials of the energy that could come from it. . . . A media campaign, "nuclear bombs are bad, but nuclear power for peace is good," began. I was an impressionable teenager who absorbed the message. Then, there was a children's show where the superhero was named "Atom," and his younger sister was "Uran-chan" [from uranium]. The first nuclear reactor went online in 1966 when I was in high school. It was the beginning of the "nuclear power era."
>
> (Koide & Sataka, 2012, p. 38)[3]

According to Koide, developing nuclear power technology became a national project and, by instituting nuclear engineering majors, seven former imperial universities effectively became the nurseries for nuclear power experts. As a high school student, Koide was convinced that his life purpose is to advance nuclear energy for human good and went on to study nuclear engineering (Koide, 2012; Koide & Sataka, 2012).

In addition to the newspaper campaigns, popular culture served as an important site for linking *genshiryoku* to righteousness. In the previous quote, Koide referenced a children's show. Indeed, many popular action and animated television shows for children and teenagers in the late 1960s to the 1980s, including *Tetsuwan Atom* (1952), *Doraemon* (1969), *Kamen Rider Series* (1971), *Mazinger Z* (1972), and *Kido Senshi Gandamu* (1981), all had heroes that relied on *genshiryoku* as their energy source. They had super-compact nuclear reactors embedded in their bodies, which provided them with unlimited energy. Regardless of the intention of the creators, these shows did not simply domesticate and naturalize *genshiryoku;* they made it a hero. The positive representation of nuclear power in the television shows, according to psychiatrist Tamaki Saito (2012), played a role in discouraging opposition to nuclear power.

While Koide and a handful of nuclear power experts later came to realize that *genshiryoku* cannot be separated from *kaku* and turned to oppose the use of nuclear power, the separation between the two words became a widespread, naturalized practice among the public in Japan. Naming nuclear power as *genshiryoku* made it possible for the public to imagine it as essentially different from *kaku*. As we see in the next sections, a few powerful articulations helped to saturate the Japanese society with the fundamental goodness and necessity of *genshiryoku*.

Absolutely safe

In the historical backdrop of Hiroshima and Nagasaki bombings and *Daigo Fukuryu-maru* incident, the promise of absolute safety was particularly critical in convincing the public of the goodness of *genshiryoku*. The pronuclear camp concentrated their effort to convey and reinforce the message that nuclear power plants are fail-proof – an articulation that came to be called *anzen shinwa* ("the safety myth"). Out of all the articulations that comprise the hegemony of nuclear power, this had been the one most underscored, but it was severely damaged as a result of the Fukushima disaster. After the disaster, newspapers, blogs, radio, and television all discussed the breakdown of *anzen shinwa* and sought to explain why this myth had been born and accepted by the public.

In an online interview (Horiuchi, 2012), Kazuto Suzuki, a public policy professor who coauthored an investigative report of the Fukushima nuclear accident, argued that the nuclear power industry insisted that nuclear power is safe in order to gain public acceptance, but everyone is responsible for the creation of the safety myth:

> Because the [antinuclear] movement opposed nuclear power, the pronuclear side had no choice but to emphasize the safety of nuclear power. I think it is this tension between the two opposing sides that escalated the safety myth. The local residents around nuclear power plants played a role, too. They wanted affirmation that it is absolutely safe before they agreed to the plant construction. Japanese citizens as a whole, too, abstractly accepted the idea that nuclear power is safe. The safety myth, in short, was created by both sides – the side that made nuclear power accident risks a taboo topic, and the pronuclear side that sought profits – to produce social acceptance.

The safety myth was widespread, and the majority of the citizens did not question the safety of nuclear power. The antinuclear movement served as an antagonism to the hegemonic insistence on safety. However, it is problematic to assume, as Suzuki appears to do, that both sides had equal power and choices and were thus equally responsible for the myth. The pronuclear bloc had vast financial, political, and media capital to control the discourse around nuclear power and suppress their opposition.

Public acceptance targeting children and women

Anzen shinwa was aggressively promoted through a number of strategies of what public relations professionals call "PA" or "public acceptance." PA is a key aspect of an industry like nuclear power whose properties are controversial. When PA is successful, the public accepts the controversial product, policy, or practice that may not be in their best interest. In a radio interview, Katsuto Uchihashi (2011), an economics commentator familiar with Japan's nuclear power history, stated that the PA of nuclear power was attained through three main strategies. First, the Federation of Electric Power Companies (FEPC) of Japan aggressively and frequently published statements to oppose unfavorable media coverage of nuclear power. Second, from elementary school to high school, pronuclear materials were taught as part of a variety of subjects such as social studies and science, and "correct" understanding of the materials influenced student grades. Finally, celebrity intellectuals were often hired to promote the idea that nuclear power is safe and reliable.

Uchihashi's interview was short and lacked specific examples of PA, but support for his argument is abundant. As will be discussed later, up until the Fukushima disaster, celebrities as well as children were frequently used in television commercials to promote nuclear energy as ordinary, familiar, and safe. But particularly significant is the PA effort directed toward children such as interactive theme parks and museums. For example, the high-tech Ibaraki Science Museum of Atomic Energy in Ibaraki Prefecture is free and open to the public. It houses a variety of interactive rooms where visitors can learn about the basics and applications of nuclear power. Notably, the museum features Albert Einstein. In "Einstein Square," modeled after Bern, Switzerland, where Einstein lived, visitors can experientially learn about the basics of nuclear power.

School children and families may also visit nuclear power plants, for the plants have public relations facilities that are open to school fieldtrips and the public. Notably, the facilities commonly use characters well known to children. For example, the energy museum within Fukushima Daini Nuclear Power Plant used Totoro, a famous creature from Miyao Hayazaki's movie, *Totoro*, to appeal to children to visit the museum that teaches the safety and reliability of nuclear power.[4] Similarly, the Shika Nuclear Power Plant adopted "Alice in Wonderland" to attract families and children. While these characters do not directly introduce visitors to nuclear power, their sheer presence hails visitors to not only come and learn about nuclear power but make it ordinary, familiar, and likable. Moreover, interactive games and animated videos educate children. An infamous example of a video is "Pluto-kun" (named after plutonium), a boy character in an 11-minute animation produced by JAEA in 1993, which was widely distributed to nuclear power plants. Pluto-kun looks human but is made of plutonium. He introduces peaceful use of plutonium and explains that, although "bad people" say he is dangerous, he won't hurt anyone even if you drink him. With a childish and sometimes sad voice and ordinary boy appearance, he entices viewers to believe him and sympathize with him.[5]

Socialization of children to nuclear power also occurred through a nationwide poster contest cosponsored by the METI and the Ministry of Education, Culture, Science and Technology (MEXT). Held annually since 1994 until the Fukushima disaster, the contest encouraged school children to learn about nuclear power. Although the aim of the contest is described as enhancement of "understanding about nuclear energy and radiation" (MEXT), the competition hailed favorable representations of nuclear power. The last call in 2010, for example, encouraged submitters to consult nine "hints" listed in the call. These "hints," presented as facts, included such statements as "radiation existed from the very beginning of the Earth" and "nuclear power plants ensure safety by using five layers of walls." In 2010, about 6,900 posters were entered for the competition ("Genshiryoku," 2011).[6] The theme of safety and reliability was pervasive in the submitted posters (METI), for the message of "nuclear is safe" was implied in many of the nine "hints" that applicants were encouraged to incorporate in their posters. One of the two winning posters, created by a 12-year-old, was a picture of nuclear waste soundly sleeping deep underground, while above the ground two girls smile and wave happily on a green meadow full of flowers. The other winner, a 16-year-old, drew a colorful picture of the Earth full of office buildings, houses, and trains. In the middle of the Earth is a large light bulb made of a flower bouquet. The words in the center read: "Nuclear power brings the future."

Williams (1977) argued that socialization, including education, inevitably ties necessary learning to "a selected range of meanings, values, and practices which, in the very closeness of their association with necessary learning, constitute the real foundation of the hegemonic" (p. 117). The contest contributed to shaping that foundation by rewarding artistic representations that reflected "correct" understanding of nuclear power. It is also noteworthy that the contest was advertised widely through schools. The competition awarded not only individual poster artists but also schools that most actively produced posters. Schools, in essence, served as a willing apparatus for domesticating and naturalizing (Deetz, 1992; Schiappa, 1989) nuclear power.

In addition to children, much of PA was directed toward women. For example, when the IAEA held the Public Information Seminar on Nuclear

Energy for a Better Life in Seoul, Korea, in October, 1993, representatives of the Japanese government, Yamada and Iguchi, presented on the importance of class-specific PA efforts and identified women as a critical class because an opinion poll showed greater uneasiness about nuclear power safety among women and also because women have considerable influence over their children's nuclear perception. Accordingly, they reported, the MITI of Japan developed magazine-based publicity, power plant visits, and lectures and meetings all designed for women.

But women became more than consumers of the PA efforts; they became a vehicle for promoting the message of safety. For example, the Japan Electrical Manufacturers' Association (JEMA)[7] established the Nuclear Power PA Women's Subcommittee to seek "accessible, women-friendly public relations that ease the public's image of nuclear power as something difficult, dangerous, and insular" ("Introducing JEMA," 2010).[8] The subcommittee's work included a variety of public relations activities such as organizing tours of nuclear power facilities and holding informal discussions with the public. They also designed annual public posters that emphasized the utmost effort power companies put in keeping nuclear facilities safe.

Identification, as Burke (1969) famously argued, serves as powerful rhetoric that helps to shape attitudes and behaviors. Acceptance of a message is more likely when it comes from someone perceived to be "one of us." The words of the former director of the Monju Nuclear Power Plant, Saburo Kikuchi, illustrate women's role in establishing identification:

> Many women are employed at Monju. The residents in the local community say that seeing these women getting on the bus everyday with no worry makes them feel very much at ease. Going to work at a nuclear plant is just part of ordinary living. That's what makes them feel at ease.
>
> (Kikuchi interviewed by Horiuchi, 2012)

Hegemony lives on such self-identification (Williams, 1977). Women, insiders in the local community, working at the nuclear power plant became a powerful tool for not only establishing the women's identification with nuclear power but also suturing nuclear power to everyday reality, making it a part of the community. It domesticates nuclear (Schiappa, 1989). The same logic perhaps underlies JEMA's use of a women-only PA subcommittee and the notable use of young women as receptionists and guides at nuclear power plant public relations offices and museums.

Suppression of opposition

The safety myth was also supported through suppression of oppositional voices. Disqualification or exclusion of opposing voice (Deetz, 1992) was a common strategy in this discursive control. In the same radio interview mentioned earlier, Uchihashi (2011) spoke about a public hearing that was held before the construction of the second reactor at the Shimane Nuclear Power Plant in the early 1980s. He remembers a mother of two children heartrendingly asking the representatives from the NSC (now the Nuclear Power Safety Authority), "If an accident happens at the plant, how are we supposed to escape? Are we expected to swim across Shinjiko? Why don't your safety criteria include an evacuation plan?" The representatives simply replied that their job at the hearing is to listen to residents' concerns and that they would not be answering any questions. The concerns were never addressed. Uchihashi has seen similar patterns over and again where those who oppose nuclear power have been ignored or otherwise ridiculed, and thus disqualified (Deetz, 1992) as being archaic and ignorant by the nuclear power industry and pronuclear scientists and scholars.

Exclusion of opposing voices was also evident after the 1986 Chernobyl accident. As noted in Chapter 2, following the accident, large-scale antinuclear demonstrations took place in Japan, and 3.6 million signatures demanding termination of nuclear power were collected and submitted to the National Diet (personal communication, 2012, September 13). Significantly, no single major media outlet covered the story of this massive and organized effort of the citizens, according to Aileen Miyoko Smith, director of Green Action (A. M. Smith, personal communication, September 13, 2012).[9] Faced with this antagonism, the government not only ignored the signatures but rearticulated the situation. Then Prime Minister Yasuhiko Nakasone declared at a congressional hearing about the accident that the design of Japan's nuclear plants is completely different from that of the Soviet Union's and that the safety of Japan's nuclear power plants has been proven. He defined the Chernobyl accident not as a nuclear power accident but as *the Soviet Union's* accident ("Genpatsu kokka," 2011). Faced with the antagonism that attempted to separate nuclear power from the safety myth, Nakasone redefined the situation to uphold the myth. Only four years after Chernobyl, an opinion poll showed that about 50% of the participants supported the increase of nuclear power and an additional 30% supported the continuation of the existing level of reliance on nuclear power (Yamada & Iguchi, 1993). It is unclear how influential Nakasone's redefinition was, but the poll suggests the tenacious link between nuclear power and safety.

Can't live without it

Public endorsement requires more than naming nuclear power as good and safe. The public must understand that nuclear power is essential to their lives. Accordingly, much of pronuclear discourse has focused on articulating this essentiality. Two themes comprise this articulation: energy independence and high demand for electricity.

Energy independence

Linking nuclear power to energy independence emerged in the wake of the oil shock in 1973. Before then, mass media had mixed responses to nuclear power. The oil shock presented an opportunity for the nuclear power industry to sell nuclear power as the key energy source to replace oil. The FEPC of Japan established a nuclear power public relations committee in 1974 to promote nuclear power through newspapers. Power companies emerged as major sponsors of newspapers in the backdrop of economic recession resulting from the oil shock. From 1974 until the Fukushima disaster, major national newspapers regularly ran stories and advertisements of nuclear power ("Kono kuni," 2012).

Since the mid-2000s, the nuclear power industry began to use television commercials as well. Using celebrities to promote products is a common practice in Japan's advertising industry, and nuclear power is no exception. For example, between 2008 and 2011, KEPCO used a popular pro-baseball coach (Senichi Hoshino) in a commercial introducing *pluthermal*, a program to reuse reprocessed plutonium as MOX fuel. The 30-second commercial shows Hoshino playing catch with two boys on a beach near a nuclear power plant. His fatherly voice says, "Japan's natural resource is scarce. We need to think about children's future." Then, a gentle female voice and a pie graph alert the audience to the statistics that Japan must import 96% of its energy without nuclear power. The female voice continues that KEPCO is working hard to actualize the *pluthermal* program to reduce waste. Excluded from the commercial is the fact that the *pluthermal* program was supposed to start in 1999 but has been postponed due to a number of safety problems and opposition from local communities. Playing catch, the beach, the fatherly baseball coach, and the soothing woman's voice, coupled with the verbal messages and statistics, all work to domesticate (Schiappa, 1989) and naturalize (Deetz, 1992) nuclear power as necessary and safe, while foreclosing undesirable representations of nuclear power.

High demand for electricity

The need for energy independence through nuclear energy is further capitalized by the argument that nuclear power is vital for meeting the electricity demands in modern life. This was the argument used by KEPCO and endorsed by the Japanese government for the resumption of the third and fourth reactors at Ohi Nuclear Power Plant in Fukui Prefecture in July 2012. This ended the two-month nuclear-free period that Japan saw for the first time since the 1970s. Soon after the restart of the Ohi third reactor, KEPC (2012) issued a press release, stating that, with the successful restart of the reactor, customers now only need to save 10% of electricity and that, without the restart of the fourth reactor, a safe and consistent supply of electricity is not achievable. Similar arguments were made by other companies; all major electric power companies received government approval to raise their electricity bills by 8–10% on the ground that they have to rely on costly imported oil and other fossil fuels to make up the electricity shortage that resulted from inactivity of nuclear reactors.

The public is thus presented with the irrefutable claim that their modern lives are not sustainable without nuclear power because natural energy sources are scarce. The claim frames nuclear power as a savior and sutures it to the existing structure – dependence on electricity. This claim, however, presumes that the current rate of electricity consumption must be maintained. The actual electricity usage in 2012 debunks this assumption. When the government called for energy conservation for the month of July 2012, the regions covered by KEPCO used 11% less than they did in 2010, a year with a very hot summer, as a result of conservation efforts by both commercial and private sectors ("Ōi Genpatsu," 2012). For the summer of 2012, KEPCO's own report showed that the peak use was 26.82 million kW, far below the maximum capacity of more than 30 million kW ("Ōi Genpatsu," 2012). This maximum capacity included 2.36 million kW generated from the two Ōhi reactors, but the peak use was still below the maximum capacity *without* the Ohi reactors' contribution. The restart of Ōi was, in short, unnecessary from the supply-demand standpoint.

The thesis of the high demand for electricity also leaves out yet another detail; the demand is not essential but was constructed. The heavy reliance on electricity in the first place was fostered and naturalized through extensive and persistent campaigns by the nuclear industry, with the help of mainstream media, including the newspapers that had been more cautious or critical about nuclear power before the oil shock. For example, *Mainichi Shimbun*, which had maintained a rather neutral stance, printed a larger advertisement of FEPC for eight months in 1976, declaring that nuclear power is the only viable energy source to replace oil ("Kono kuni," 2012). The recurrent advertisement included a photograph of a woman cooking, clearly linking basic daily activities to the necessity of nuclear power.

More recently, from the mid-2000s until the Fukushima disaster, power companies aggressively promoted *ōru denka* ("all electric") campaigns to encourage consumers to go "all electric" on their appliances. For example, TEPCO ran a series of commercials that depicted a young family – a mother, a father, and a pre-teen son – excited about their new electric appliances and water heater. All this is humorously done, selling the idea that "all electric" is modern and fun and makes a happy family. The *ōru denka* campaigns paid off. According to TEPCO, the number of "all electric" households went from 13,000 in 2002 to 456,000 in 2008 and to 855,000 in 2010 ("Ōru denka," 2011). The exponential increase from 2008 to 2010 led to the increase of 2 million kW – equivalent to the volume of electricity produced by two nuclear reactors. Thus, the consumers of *ōru denka* campaigns became willing agents in promoting the heavy reliance on electricity and, by extension, the use of nuclear energy. Ironically, it was these all-electric households that suffered the most when the power conservation was enforced after the Fukushima nuclear disaster.

Green and clean

Nuclear energy as a solution to global warming is another articulation that has been widely used to promote nuclear power. Nuclear power as "green and clean" became a powerful slogan since the United Nations Framework Convention on Climate Change (UNFCCC), or the Earth Summit, was held in Rio de Janeiro in 1992 and was attended by 189 nations' representatives. It was at this convention that CO_2 was declared as a serious contributor to global warming. Just as the 1973 oil shock presented an opportunity to promote nuclear power for energy independence, this declaration – and later the Kyoto Protocol of 1997 – served as a welcome prompt for declaring nuclear energy as the most desirable green alternative to oil.

The nuclear industry aggressively used a number of media and public relations venues to promote nuclear power as the most viable green energy. Television commercials, again, were one of the popular channels. Power companies and the Nuclear Waste Management Organization of Japan (NUMO) aired television commercials conveying that nuclear energy does not produce CO_2 and therefore is clean and kind to the environment. Many of these commercials featured or referenced children to remind the audience that supporting nuclear energy also means that you are protecting children and future generations. For example, TEPCO frequently ran a commercial featuring Hitoshi Kusano, a veteran host of news programs. In one commercial, Kusano sits on a deck with a girl who is making vegetable juice in a blender. Kusano says Japan's electricity is much like the vegetable juice, because it comes from a variety of sources such as solar, hydro, and oil. Pictures of these power facilities flash for a few seconds. Then Kusano quickly turns to nuclear power, to which the rest of the commercial is devoted: "Nuclear power does not produce CO_2. A nuclear reactor that generates 1 million kW can reduce CO_2 as much as the amount of all the trees in the Kanto region can absorb in a year." The very last screen shows TEPCO's name and an invitation: "let's switch to a low CO_2 life style."

Another outlet is public relations posters that were widely displayed at a variety of public spaces. Every year until the Fukushima nuclear disaster, JEMA published a poster designed by the aforementioned women's PA subcommittee. Many of these posters emphasized the earth friendliness of nuclear power. The 2004 poster, for example, features a girl standing with her arms stretch upward. Above her is the Earth, the size of a big beach ball. The heading says, "We need Genshiryoku" ("we need" is in English). Under the heading, the text reads:

> Blue Earth and green land, the vast history of life. The Earth seen from the universe is a concentration of miracles. The Earth is calling

for ecological energy. Nuclear power does not produce CO_2. Nuclear power is the energy of the future.

This poster and others like it declare nuclear power as ecological and the most suitable energy source that protects the past, present, and future. While these posters may be produced for adult audiences, the government-sponsored poster contest discussed earlier served as an important tool for educating children about the earth friendliness of nuclear power. As stated, the "hints" about nuclear power that children are encouraged to incorporate in their posters included such information as "nuclear power is kind to the Earth because it does not produce CO_2 while generating electricity" and "uranium used by nuclear power plants is recyclable."

As shown, in the pronuclear discourse, "green energy" is articulated exclusively in terms of no production of CO_2 *during electricity generation*. This is the recurrent message across a range of representations of nuclear power from television commercials to public relations posters. Left out from the articulation is what occurs before and after electricity generation. Environmental groups and nuclear power critics everywhere have all pointed out, and the nuclear industry itself has admitted, that mining, import and transportation of uranium as well as construction of nuclear power plants all rely on fossil fuels and produce CO_2. Nuclear reactors also use a massive volume of water for cooling and purge hot water back into the oceans, altering the ocean ecosystem. Moreover and significantly, the radioactive waste from nuclear power is not biodegradable and remains radioactive for thousands of years. The hegemonic articulation of green energy, however, naturalizes exclusive focus on CO_2. By way of naturalization and topical avoidance, it creates closure around the discourse (Deetz, 1992) of nuclear energy and ecology by silencing other environmental consequences that accompany nuclear power technology.

Conclusion: reflections on hegemony

Japan's nuclear power achieved its hegemonic status by penetrating the everyday life of citizens through public relations campaigns, mass media, and education. Using a variety of discursive practices, pronuclear discourse persistently fixed the identity of nuclear power as virtuous, absolutely safe, necessary, and environmentally friendly. Each practice may seem innocuous, ordinary, and unrelated to others. Yet collectively such practices are constitutive of a discourse that promoted and naturalized nuclear power. While I presented the four articulations separately, there were recurrent themes across the articulations. In this section, I discuss these themes in Japan's pronuclear discourse to reflect on the concept of hegemony.

First, across the articulations, pronuclear public relations efforts consistently focused on socialization of children. The pronuclear public relations campaigns used children as instruments of a larger social formation and policies regarding energy and modern life. This is especially salient, given that understanding nuclear power requires highly technical knowledge even difficult for educated adults. To what extent can children understand and evaluate the information presented to them? Perhaps precisely because children lack such ability, they serve as ideal instruments for advancing hegemonic discourse. Sha and Meyer (2002) note that, while there is abundance of studies of propaganda or communication campaigns that seek to alter children's attitudes and behaviors, very little research is available on the use of communication campaigns that use children as instruments of social change. To my knowledge, this remains the case. Any effort to understand and destabilize hegemony may benefit from examining the ways in which children become instruments of social formation and social change for better or worse.

Second, in advancing the four articulations, the pronuclear bloc has had tremendous power in controlling the discourse of nuclear energy. This concentration of power cannot be quite explained through a commonly embraced understanding of hegemony and power. Noting their departure from Gramsci's view of hegemony, Laclau and Mouffe (1985/2001) argued that hegemony must be understood in terms of "democratic struggles" where there is no single hegemonic center but only multiple nodal points of power. Of course, this is also the way Foucault (1990) saw power: power is everywhere, it comes from everywhere, and it is in every relation. Rejecting a totalizing, top-down view of power, Foucault and Laclau and Mouffe alike encourage inquiry into everyday practices of power that contribute to hegemony.

While the multimodality view of power is now canonical in social inquiry, DeLuca (2011) questions its applicability to today's world:

> Sure, power is exercised from innumerable points, and ExxonMobil is one such point, and my neighbor is another point, but to not differentiate among points and note the enormous privileging of certain points is to render one's analysis irrelevant. There are centers of power, and they matter.
>
> (p. 90)

DeLuca's observation rings true for Japan's pronuclear discourse we have seen in this and last chapters. The pronuclear bloc had considerable power in creating and controlling the discourse of nuclear power through the public relations, media, and education campaigns to produce a regime of truth (Foucault, 1972/2010). The citizens, participating in the discourse as consumers and messengers, are hardly equivalent nodal points. Social inquiry into hegemony must consider how this concentration is maintained and challenged. Can the concentration of power be undermined? While we may draw a blank, considering the examples discussed in this and last chapters, there were spaces for democratic struggle in Japan's nuclear power. In the next chapter, we will turn to these spaces.

Notes

1 With permission, a large section of this chapter was taken from Kinefuchi, E. (2015). Nuclear power for good: Articulations in Japan's nuclear power hegemony. *Communication, Culture, and Critique, 8*(3), 448–465. https://doi.org/10.1111/cccr.12092

2 *Genshiryoku mura* refers to the nuclear-industrial complex and its insularity. It includes the utilities, nuclear vendors, the National Diet (Japan's parliament), financial sectors, media, and academia that collectively promote nuclear power.

3 All direct quotes from Japanese language sources are my translations.

4 *Totoro* disappeared after a complaint was made by its creator, Hayao Miyazaki, who has been a vocal opponent of nuclear power.

5 Nuclear power plants stopped using the video a few years later.

6 The 2011 competition was cancelled and has not resumed since the Fukushima disaster.

7 According to JEMA's website, JEMA consists of major Japanese companies in the electrical industry, including power and industrial systems, home appliances and related industries.

8 *Metropolitana* is a free monthly magazine distributed throughout subway stations in Tokyo. The story is available online, but no post-Fukushima information about the subcommittee's activity is available on the Internet, including JEMA's website.

9 See more details about Aileen Miyoko Smith in Chapter 6.

References

Arima, T. (2008). *Genpatsu, Shoriki, CIA*. Tokyo: Shincho Shinsho.

Artz, L., & Murphy, B. O. (2000). *Cultural hegemony in the United States*. Thousand Oaks, CA: SAGE Publications, Inc.

Burke, K. (1969). *A grammar of motives*. Berkeley: University of California Press.

Deetz, S. (1992). *Democracy in an age of corporate colonization: Developments in communication and the politics of everyday life*. Albany: State University of New York.

DeLuca, K. M. (1999). Articulation theory: A discursive grounding for rhetorical practice. *Philosophy & Rhetoric, 32*(4), 334–348. Retrieved from www.psupress.org/journals/jnls_pr.html

DeLuca, K. M. (2011). Interrupting the world as it is: Thinking amidst the corporatocracy and in the wake of Tunisia, Egypt, and Wisconsin. *Critical Studies in Media Communication, 28*(2), 86–93. https://doi.org/10.1080/15295036.2011.572680

Doyle, J. (2011). Acclimatizing nuclear? Climate change, nuclear power and reframing of risk in the U.K. News media. *International Communication Gazette, 73*(1–2), 107–125. https://doi.org/10.1177/1748048510386744

Foucault, M. (1972/2010). *The archaeology of knowledge and the discourse on language* (A. M. Sheridan Smith, Trans.). New York: Vintage Books.

Foucault, M. (1990). *The history of sexuality, volume 1: An introduction* (R. Hurley, Trans.). New York: Vintage.

Genpatsu kokka: Nakasone Yasuhiro hen [Nuclear Power Nation: Nakasone Yasuhiro Issue]. (2011, July 21). *The Asahi Shimbun*. Retrieved from http://blog.goo.ne.jp/harumi-s_2005/e/894e9105d199310c1457db0eb7605b73

Genshiryoku posuta konkuru wo chushi . . . monkasho nado [MEXT and Its Cosponsor Cancel the Annual Nuclear Power Poster Contest]. (2011, May 11). *The*

Yomiuri Shimbun. Retrieved from www.yomiuri.co.jp/national/news/20110511-OYT1T00880.htm

Hall, S. (1985). Signification, representation, ideology: Althusser and the post-structuralist debates. *Critical Studies in Mass Communication*, 2(2), 91–114. https://doi.org/10.1080/15295038509360070

Horiuchi, A. (2012, June 19). Naze genpatsu no anzen shinwa wa umaretanoka [Why was the nuclear power safety myth born?]. *Business Media Makoto*. Retrieved from http://bizmakoto.jp/makoto/articles/1206/19/news023.html

Ibaraki Science Museum of Atomic Energy. Homepage. Retrieved from www.iba gen.or.jp/

Introducing JEMA "Nuclear Power PA Women's Subcommittee." (2010, October). *Metropolitana*. Retrieved from www.metropolitana.jp/contents/1010/nuclear.html

Kansai Electric Power Company. (2012, July 9). *Ohi 3goki no saikado ni tomonaru setsuden naiyo no minaoshi ni tsuite [Revision to power conservation upon the restart of Ohi 3rd nuclear power reactor]*. Retrieved from www.kepco.co.jp/pressre/2012/0709-2j.html

Koide, H. (2012). *Genpatsu to kenpo 9 jyo [Nuclear power and Article 9]*. Naka, Japan: Yushisha.

Koide, H., & Sataka, M. (2012). *Genpatsu to nihonjin [Nuclear energy and the Japanese]*. Tokyo: Kadokawa.

Kono kuni to genpatsu: media no kattou [This Country and Nuclear Power: The Media's Struggle]. (2012, October 22). *Mainichi Shimbun*. Retrieved from http://mainichi.jp/feature/20110311/news/20121022ddm003040191000c.html

Laclau, E., & Mouffe, C. (1985). *Hegemony and socialist strategy: Towards a radical democratic politics*. London: Verso.

The Ministry of Economy, Trade and Industry (METI). *Dai 17 kai genshiryoku postar konkuru no jyushou sakuhin no happyo nit suite [On the 17th annual poster contest winner announcement]*. Retrieved from www.meti.go.jp/press/2010 1105004/20101105004.html

The Ministry of Education, Culture, Sports, Science, and Technology (MEXT). (n.d.). *Dai 17 kai genshiryoku poster konkuru no jisshi ni tsuite [On the 17th annual poster contest]*. Retrieved from www.mext.go.jp/b_menu/houdou/22/06/1294 980.htm

Ōi Genpatsu: Saikado wa hitsuyo dattaka [Ohi Nuclear Reactor: Was the Restart Necessary?] (2012, September 3). *Mainichi Shimbun*. Retrieved from http://maini chi.jp/feature/news/20120903dde012040009000c.html

Ōru denka jyutaku, fukyu urame, genpatsu niki bun no shouhizou [The Increase in Power Usage from All Electric Households Is Equivalent to Two Nuclear Power Reactors]. (2011, March 23). *The Yomiuri Shimbun*. Retrieved from http:// www.yomiuri.co.jp/atmoney/news/20110323-OYT1T00569.htm

Saito, T. (2012). *Genpatsu izon no seishin kozo [Psychology of nuclear power dependence]*. Tokyo: Shinchosha.

Schiappa, E. (1989). The rhetoric of nukespeak. *Communication Monographs*, 56(3), 253–272. https://doi.org/10.1080/03637758909390263

Sha, B., & Meyer, K. C. (2002). Children and government propaganda. *Journal of Promotional Management*, 8(2), 63–87. https://doi.org/10.1300/J057v08n02_06

Slack, J. D. (1996). The theory and method of articulation in cultural studies. In D. Morley & K. Chen (Eds.), *Stuart Hall: Critical dialogues in cultural studies* (pp. 112–130). London: Routledge.

Slack, J. D. (2006). Communication as articulation. In G. J. Shepard, J. St. John, & T. Striphas (Eds.), *Communication as . . . perspectives on theory* (pp. 223–231). Thousand Oaks, CA: SAGE Publications, Inc.

Uchihashi, K. (2011, March 29), Genpatsu anzen shinwa wa ikani tsukuraretaka [The making of the nuclear safety myth]. *NHK Radio Daiichi.* Retrieved from www.nhk.or.jp/r-asa/business.html

Williams, R. (1977). *Marxism and literature.* Oxford: Oxford University Press.

Yamada, M., & Iguchi, T. (1993). *PA activities of nuclear power in Japan.* Retrieved from www.iaea.org/inis/collection/NCLCollectionStore/_Public/25/070/25070426.pdf

Yoshida, Y. (2007, June 23). *Kaku to genpatsu no chigai [The difference between "kaku" and "genpatsu."* Retrieved from www.yoshida-yasuhiko.com/nanp/post-90.html

5 Fighting for community
Antinuclear movements at ground zero

Why pay attention to local site fights?

The preceding chapter discussed communicative strategies that helped to hegemonize nuclear power as an indispensable technology for advancing Japan's economy and modernization. Hegemony, however, should not be taken to mean that the social process was accepted by citizens and communities blindly. On the contrary, nuclear power was met with opposition in every community sited for nuclear power plant construction. Since the first reactor went online in 1966 in Tokaimura, Ibaragi Prefecture, on average, two reactors were built yearly until the 1990s. At the time of the Fukushima accident, Japan owned 54 operating reactors and was the third largest nuclear power nation after the United States and France. These statistics may appear to suggest that the whole society was on board with the government. What is not acknowledged, however, is the number of facilities that were never built. Indeed, over 30 projects were abandoned due to local resistance (Yamaaki, 2012; Behling, Behling, Williams, & Managi, 2019). Another study (Hirabayashi, 2013) estimates that the number of abandoned projects is actually around 50 if we include the projects that were cancelled but were not yet formally declared by utilities as such. This means, for almost every reactor that was built, another reactor was stopped due to opposition.

These local resistances are at the heart of Japan's antinuclear movements. Most of the reactors that were built in the 1970s or later were additions to the existing plants, not brand-new locations. According to Baku Nishio, co-director of the CNIC, this is because local movements successfully blocked new plant projects (B. Nishio, personal communication, June 13, 2013). Each of these victories *and* ongoing resistances is extraordinary, given the sheer duration of the commitment required for fighting the government-utility complex. In 2012–2013, I visited six municipalities where site fights occurred, including Maki-machi and Kashiwazaki in Niigata, Hidaka-cho in Wakayama, Iwaishima in Yamaguchi, and Oma and Rokkasho in Aomori. In all these municipalities, the fights against the constructions of nuclear

DOI: 10.4324/9781003044222-5

power plants lasted for 20 or more years. In Iwaishima and Oma, the fights that began in the 1980s continue today.

This chapter features two of these communities, Hidaka-cho and Maki-machi, where plant construction was eventually abandoned due to steadfast opposition. The question that guided my research was simple: What factors led these movements to succeed? Nuclear power is a national policy with massive political and economic clout that could easily swallow its opposition. It is worth learning about the social practices that enabled these small rural communities to successfully fight against the enormous power. While each community's circumstance is unique, I elucidate both the uniqueness and the common threads across these cases. Pre-Fukushima antinuclear movements, particularly the local ones, are relatively unknown outside Japan – and even inside Japan – but are the ground zeros of nuclear power struggles and thus are indispensable to the understanding of Japan's nuclear power discourse. Nuclear site fights are life-altering, long-lasting major events for the communities that were targeted for a nuclear facility project. The struggles were reported by local news media but were given scant attention by national television or newspapers. In addition, as we learned in Chapter 3, national papers habitually represented local oppositions as a matter of the citizens' lack of correct understanding or disapproved local residents' direct involvement with nuclear power matters.

The stories of two communities' antinuclear struggles I share are based primarily on the interviews I conducted with community activists and the documents (e.g., newsletters) they produced and are supplemented by other secondary publications. Because the goal of the chapter is to tell the local activists' stories, I take a narrative approach to discourse. As with Walter Fisher (1985), I use narrative to mean a whole range of symbolic actions that collectively creates a coherent story that has meaning for the actors and for those who interpret it. Narrative, therefore, "comes close to capturing the experience of the world" (Fisher, 1984, pp. 14–15). While experience and storying of the experience are not identical, there is an intimate relationship between story, experience, and sense-making from a phenomenological perspective. That is, experience becomes part of consciousness through storytelling (Squire, 2013). Storytellers give meaning to their experience as they symbolically share the experience and, in doing so, their storied experience is brought into the consciousness of their listeners. Thus, the stories in the following pages are layered mediated symbolic representations. Additionally, an experience may be differently storied depending on the context in which the story is told (Squire, 2013). Then, what makes these stories trustworthy? The fidelity of narrative, Fisher (1984) argues, comes from narrative rationality where the important question is "whether the stories they experience ring true with the stories they know to be true in their lives" (Fisher, 1984, p. 8). Thus, keeping coherence in mind for narrative fidelity, I share the stories of antinuclear activism at the ground zero of site fights

Figure 5.1 Map of Japan showing two site fight locations relative to Tokyo and Fukushimna.

Source: Map created by the author.

that offer understanding of the social practices that led the activism to a successful ending.

Hidaka-cho, Wakayama

Wakayama Prefecture covers a large part of Kii Peninsula, the largest peninsula in Japan. It is home to dense forests and beautiful coastlines. It was 1967 when local newspapers reported the KEPCO's plan to build the first nuclear power plant in Wakayama Prefecture. From the 1960s to 1980s, the

Figure 5.2 The proposed nuclear power plant site, Hidaka-cho.
Source: Photograph by the author.

prefecture became a battleground over the construction of nuclear power plants. At least five locations were targeted by KEPCO as potential sites, which caused intense and long-standing conflicts within these communities between those who supported the project and those who opposed it. This chapter focuses on the one at the Oura district in Hidaka-cho where the fight was particularly arduous.

Utility's strategies for buying land and community support

The construction of a nuclear power plant typically takes five to ten years, but the project begins many years prior to the construction. It first requires environmental assessments to determine the geological and geographical suitability of the general area of interest. Once that is done and a potential site is identified, the land needs to be purchased and the consent of the local community must be acquired. Land purchase can pose a tremendous challenge to the project, for a nuclear power plant requires hundreds of acres owned often by multiple individual owners. While KEPCO pursued two sites, Ao and Oura, in Hidaka-cho, they avoided direct negotiation with landowners in both cases. In Ao, they had a lumber company purchase the land and then have its ownership transferred to them. Feeling betrayed, the landowners sued the lumber company but lost. However, the whole ordeal fortified the stakeholders of Ao – the Hiizaki Fishery Association and the district leadership – to declare their opposition to the project.[1]

As the opposition was growing in Ao, KEPCO shifted its attention to an adjacent district, Oura. By the time they approached Oura in June 1975, the utility had already secured a large piece of land for their nuclear project. This was done indirectly so that the real buyer and purpose of the sale were obscured. In 1970, the land was first purchased by the Hidaka-cho municipal government. The mayor explained that it would be used for tourism development. Within a year, however, the land was sold to a trading company associated with KEPCO (Terai, 2012).

While the land was being acquired, KEPCO worked to garner the support of the community to attain their consent to the nuclear project. To begin, the company established a site department whose sole job was to persuade the residents to support bringing a nuclear power plant to the community. From educational activities targeting children to public relations events, the site department utilized a variety of strategies to win over the residents, but the strategy that they put the most effort in was the nuclear energy study tours (Asami, 2012). The utility first took the district chief to tour their plants in Mihama and Ohi in Fukui Prefecture. Upon returning, the chief reported that the plants are safe and strengthen local economy (Terai, 2012). Over multiple years, KEPCO also actively invited district residents and city officials to the luxurious tours that it organized and paid for. Although the tours were promoted in the name of better understanding of nuclear power, they were orchestrated so the participants only learn the advantages of nuclear power (Asami, 2012). One resident, Takuya Terai, published his tour experience in a magazine, *Gijyutsu to Ningen* (Technology and human), reporting that he was bombarded with pronuclear messages – nuclear power is safe, makes life easy, and brings prosperity to local economy – and learned nothing about potential dangers and disadvantages that accompany nuclear power (Asami, 2012).

Kiyokazu Hama, a local fisherman and board member of the local fishery association, participated in several of these tours, wanting to learn how the nuclear towns faired. Mr. Hama is now deceased, but his son, Kazumi, also a fisherman and one of the leaders of the opposition movement, told me why his father got involved in the local antinuclear movement. In one of the tours, Kiyokazu took a detour and met a local fisherman. He asked what it was like to have a nuclear power plant in his village. The fisherman in return asked, "Do you have a child who would be taking over your fishery business?" When Kiyokazu confirmed that he did, the fisherman said, "Don't let them build a plant. Nothing good will come out of it." After this experience, Kiyokazu became active in building an antinuclear movement.

The conclusion that Terai and Hama drew were the opposite of what these tours intended. The tours were meant to win the hearts of the participants, and it worked for many; the tour participants saw beautiful roads and state-of-the-art buildings funded with nuclear program subsidies and visited a fish farm where carp were raised in the water used for cooling the reactors. They were pampered with gourmet food and drinks. They came back convinced; nuclear power is the necessary national policy, the safe solution for the energy demand, and the source of economic prosperity for local communities thanks to handsome subsidies (Nakanishi, 2012).

KEPCO also bought the support of local residents by persistently making home visits. Masayo Matsuura, a leader in the antinuclear movement, described to me how this occurred:

> You see KEPCO's site department employees around the town, visiting the homes of the residents who were in the opposition camp. They come to collect information, to watch their moves, and to convince them to switch sides. When they visit the same house multiple times, it is pretty obvious. Then, you learn that the folks in that house suddenly became pronuclear.

The residents did not become pronuclear simply because they were convinced by KEPCO's reasoning of why they should support their nuclear program. They switched sides because they got something in exchange. According to Kazumi Hama, the gift came in different shapes. It could be money. It could be securing employment for their sons and daughters with KEPCO or affiliated companies. Masayo had a similar story. One of the women leaders in the movement had a child who was about to go to college. KEPCO offered to pay for college. Other women who had children looking for employment were told that KEPCO would find them jobs. When another woman's husband was hospitalized, KEPCO offered to chauffer her between home and the hospital. These strategies were all effective, according to Kazumi; one by one, he saw his neighbors and friends turned into KEPCO supporters.

Unraveled community

KEPCO's nuclear power plant project tore apart the community for two decades. Families and relatives that used to be close were divided. Although that is now over 30 years ago, Shizue Suzuki vividly recalled how everyone was miserable:

> It was tough. People were divided into either pronuclear or antinuclear side. You don't shop at that store because its owner is pronuclear. The people who worked for the town government were all pronuclear. . . . Families were sharply divided and even important family events like funerals couldn't bring them together. The gulf between the two sides was so deep that the tension and awkwardness lasted long after the nuclear project was cancelled.

This was also the case within the fishery association that is normally known for a tight-knit brotherhood. For years, the fishermen in Hidaka were forced to take sides. Kazumi described the situation as follows:

> We, fishermen, rely on each other. Rough seas and extreme weathers. We must help each other. Some years ago, a fisherman fell from his boat and died. As an association, we all looked for the man for a week, making a huge net and pulling it with large ships. Together. But nuclear power divided us and made us fight unnecessarily. When an association member on the pronuclear side built a new ship, those on the antiulcer side were not invited to the ceremony. Nuclear power split families, siblings, and relatives. When a daughter has a wedding, her relatives should be there to celebrate her big day. But, if you were on the other side of the nuclear power fight, you didn't get invited.

The fishermen in both sides collided every time when there were meetings to discuss whether the association permits KEPCO to conduct a coastal survey. Verbal fights sometimes escalated into physical fights.

Fight to preserve community

In the end, what led the opposition to prevail? Several factors seem to have made it possible. First, KEPCO was unable to obtain the fishery association's consent to their coastal survey, a prerequisite for nuclear power plant construction. The plant would extract a large volume of water daily to cool reactors and dump it back to the ocean. This practice undermines the ocean ecology because the spent water is extremely hot, mildly radioactive, and contains chemicals. Consent means accepting these consequences and waving fishing rights to the substantial waters around the nuclear power plant. Hidaka's association never gave the consent. This does not mean that the association was united. In fact, KEPCO's strategies were working, and many association members jumped ship to the pronuclear side. All but 3 of the 14 board members became supporters. The key variable (and the second factor) that prevented the association from giving the permission was the dedicated, unwavering few leaders who refused to give into the majority. Kazumi was one of them. His father, who initially was active in the antinuclear movement and supported Kazumi's antinuclear activities, reprimanded him for single-mindedly devoting himself to the activism and neglecting his family when other fishermen were earning a lot of money. It is true that Kazimi's wife had to run their inn without his help during this period, but this was a fight that could not wait, whereas he could always make money once the ocean, which provides for fishermen, is safe from nuclear power. Kazmi and other members in the opposition camp worked tirelessly to put an antinuclear member into the association chair position. This proved to be a losing battle, but they did not give in. Kazumi described the struggle:

> The chair gave in to the lure of money from KEPCO. And he was not the first. The chair before was also antinuclear, so we supported him to become the chair. We helped him to win a seat in the town council, too. We worked day and night to make sure that the antinuclear side gets represented. But six months later, he flipped. You hear through the grapevine that this person got paid 5,000,000 yen, that person got paid 10,000,000 yen. KEPCO used money as a weapon, and it worked, because the people we thought were our comrades kept betraying us. But we never gave up. When an association meeting was coming up, we stayed up all night to prepare for the meeting. In case the issue of supporting the nuclear program comes up, we needed to be always ready. We would call Koide-sensei and Imanaka-sensei, and they would come to help.

Despite the repeated betrayals and being in a clear minority, they kept up their fight. A key to their survival was being thoroughly prepared to defend their position, and this groundwork was aided by nuclear scientists at the Kyoto University Research Reactor Institute (KURRI) in Kumatori-cho, Osaka. Support from nuclear scientists may sound odd because the goal of the institute was to foster research and education in nuclear energy and radiation application. A reasonable assumption may be that the faculty is entirely in favor of nuclear energy. However, among some 80 members of the KURRI were six men who took a critical stance on the use of nuclear energy. The six rebel nuclear scientists, later known as *Kumatori rokunin-shu* (Kumatori Six), provided expert support to the antinuclear movement in Oura.[2] Kazumi would call these sensei, and they would come to help even late at night. They together discussed tactics until the next morning. With the help of these experts, the opposition side steadfastly averted the association meetings from being taken over by the pronuclear agenda.

Oura's antinuclear movement, however, did not just exist in the fishery association. Other residents were just as involved, and the support from the surrounding communities and towns and even from outside Wakayama was immensely important. They included doctors, mothers, fathers, teachers, and people from other prefectures who were fighting the same fight. They provided concerted moral and physical support by participating in demonstrations, creating and distributing flyers, visiting residents door to door to collect signatures, and pressing town officials. Some 30 doctors came together and made flyers describing the harms that result from radiation exposure and inserted one in every copy of the local newspaper (Terai, 2012). And women rose up. Shizue Suzuki, one of the women leaders in the movement and then an elementary school teacher, recounted her experience and stressed the centrality of women and mothers in the movement:

> Initially, we didn't know what nuclear power was, but many of us residents were in favor of it because we were told that they would give us a lot of money, and they would build a new school. The existing school was really old and dilapidated. But then we learned the truth about nuclear power and began to protest. Women were strong. Fishermen tend to be quiet, because they deal with fish and ocean. But women were at the forefront of the antinuclear movement.

Shizue led the formation of *genpatsu ni hantai suru onna no kai* (the organization of women who oppose nuclear power) and mobilized mothers. She recalled the time when a pronuclear national senator visited the town, trying to move the nuclear project forward. A group of mothers surrounded the senator's car and even put their feet in the car, telling him to back off. The senator could not get out of the car and eventually left without meeting the town leaders. Women also strategically organized to collect thousands of signatures to oppose nuclear power while the mayor was out of town for "study tours" to other nuclear towns.

Masayo Matsuura, a leader in the opposition movement, confirmed the important role that women played in the movement. Masayo first learned about KEPCO's plan to build nuclear power plants in Wakayama when she saw the news of the TMI accident in the media. She realized the danger of nuclear power and wanted to learn more about it. She began to hold study sessions with others through a learning group, *Penpen gusa*, which she was co-organizing in Wakayama City. Through these sessions and visiting the towns where nuclear power site fights were taking place, she became increasingly involved in the movement. She helped to form *Kokyo wo mamoru onna no kai* (the organization of women who protect their hometown) in 1987. This was an alliance of ten small antinuclear women's groups that emerged in different parts of Wakayama. It is noteworthy that most women in these groups were married to men with non-corporate jobs (e.g., doctors). In Hidaka-cho, the antinuclear activities were closely watched by the police and were reported to KEPCO. And KEPCO, as one of the giant conglomerates in the country, exerted far-reaching power to make corporate men vulnerable even if they don't work for KEPCO. Masayo witnessed women leaving the groups after their husbands learned about their involvement with the groups. The women who stayed in the groups persevered, however. They held events in their own local communities to learn about nuclear power. They created flyers and placed them in local newspapers. They fundraised. They went to the town and prefectural governments (which were almost all pronuclear) to voice their objection. They visited each residence in Hidaka-cho to talk to people face to face. These grassroots efforts were critical in keeping the oppositional voices alive in the sea of the pronuclear voices that had unlimited financial resources at their disposal and the police force on their side.

In June 1984, the fishery association met to discuss once again whether to permit the coastal survey. At that time, the association was facing a grave deficit due to multiple fraudulent loans. KEPCO offered 300 million yen to the association in exchange for the survey.[3] The meeting was convened to discuss this matter. Soon after the chair's opening speech, a fight erupted over the presence of prefectural officials in the meeting, and it escalated into physical fights. To stop the fight, the chief declared the meeting cancelled and announced his resignation from the chairmanship. After this contentious episode, the coastal survey issue surfaced several times but never got resolved. Meanwhile, the Chernobyl accident occurred in 1986 and made the residents fearful of nuclear power.

The survey issue resurfaced in 1988, and the association convened to take a vote. One difference between this time and the previous meetings about this matter was that the association leader and KEPCO already met to discuss the price of allowing the survey – 670 million yen.[4] This meeting, too, led to verbal fights and then became physical to the point that the chief was hit on the face. About 30 officers from the riot police arrived to calm everyone down. After the chaos that lasted three hours, the chair declared that the association would drop the topic and that the whole association leadership would resign. This meant no vote, and no vote meant no survey. Two years later, the association board officially avowed its commitment to the "no survey" decision. In 2005, the national government removed Hidaka-cho from the list of sites for power source development.

What eventually made the fishery association chair give in? In his interview with *The Asahi Shimbun*, the chair expressed regrets for the trouble the whole situation caused the prefecture and Hidaka-cho and stressed that the bottom line for him was fishermen's well-being; he did not want to see the association members fight anymore, and the nuclear project needed to be scrapped so the members can come together again (Terai, 2012). The chair's remark suggests that, despite the brutal fights that spanned over 17 years, the community integrity to protect what's important for the community eventually prevailed over money. Trust, relationships, future generations, and the ocean were the core of the community and what Kazumi, Shizuka, Masayo, and others in Hidaka's antinuclear movement were fighting for. Kazumi's words capture what nuclear power means for him and why he fought two decades to keep it out of his town:

> In a rural community, we are all connected to each other like a net, and the net is essential to our existence. If the net is tight, there won't be a nuclear power plant. But they come in and try to cut it into pieces. That's how a plant gets built. Even a fishery association chair who opposed the plant gets turned around. It tears up your town and destroys the trust you had with each other. But it doesn't hurt KEPCO at all. They don't

have a sense of what it means to be human. That is how you build a nuclear power plant.

Maki-machi, Niigata

Around the time the KEPCO came to Hidaka-cho, the Tohoku Electric Power Company (Tohoku Denryoku) had a plan for Maki-machi, Niigata Prefecture. Before it was merged into Niigata City in 2005, Maki was an independent municipality southwest of the city. In 1969, *Niigata Nippo*, a local newspaper, published an exposé that Tohoku Denryoku was planning to build a nuclear power plant in Maki. Just as other towns targeted for nuclear power plant construction, Maki was a quiet, coastal and agricultural town. But these towns had something else in common. They were economically struggling and politically conservative. Maki was no exception. If seven out of ten farmers were pronuclear, that was because they were unable to earn sufficient living solely from rice farming. Farmers hoped a power plant would bring stable jobs to their families, and small business owners, too, saw it as an opportunity to boost the town's economy (Ito, Watanabe, Matsui, & Sugihara, 2005). Mrs. Kuwabara, a Maki resident noted, "Nishikan Ward [in which Maki is located] was and is still conservative. If you do something against the norm, you will be marginalized. Gender stratified, too. When a woman ran for the town council, it was a very big deal and became national news." Typical of rural towns in Japan, Maki's political culture was dominated by the conservative, pronuclear Liberal Democratic Party. Traditionally voting decisions were made by the head of a household, and women were expected to stay away from politics. This meant that, as practice, husbands assumed power to influence their wives' voting behavior. With this cultural climate as a backdrop, the town council passed a resolution in 1977 to support the construction of a nuclear power plant.

This resolution and political conservatism alone were a significant hurdle for the residents who opposed the nuclear project, but a serious blow came in 1980 when the Maki Fisheries Cooperative Association gave in after maintaining an opposing stance for years. Despite the strong objections from some members, the association agreed to the plant construction in exchange for 3,900 million yen.[5] As we saw in Hidaka-cho, the local fishery association's agreement often holds the most critical key to the question of which side of a site fight wins. Without it, the project will not proceed even when the town council, the mayor, and the governor all endorse the project. With the fishery association's consent, the construction of the Maki Nuclear Plant officially became part of the nation's Basic Electric Energy Development Plan. That is, the government determined that this plant would be critical for Japan's energy security. In spite of this giant step toward construction, the opposition managed to eventually reverse the course. In the following, I discuss the development of Maki's antinuclear movement and key contributors to this remarkable victory.

The birth of Maki's antinuclear movement

According to Mrs. Kuwabara, a local activist and retired high school teacher, Maki's antinuclear movement began when a local plumber, Torao Endo, gathered his old friends from junior high school to learn more about nuclear power after reading the 1969 exposé. This small circle of friends met at the town library to share what they learned. Soon, they realized that, if they were to have any chance of fighting, they must involve the residents of Gokahama, a village adjacent to Kakumi, a proposed construction site. Six months after the exposé, they began going around door to door every weekend to get the villagers involved. After three months of persistent visits, the residents who initially refused to hear Endo and his friends began to invite them for a tea. Overtime, they felt as though they were visiting their own home (Ito et al., 2005). Endo believes that he was able to build trust most likely because the villagers saw him as a fellow resident whom they sometimes see around the town doing plumbing work or being with his pregnant wife (Ito et al., 2005). Eventually, the villagers rose up and formed an association aiming to protect Gokahama (*Gokahama wo mamoru kai*). This was August 1971, three months after Tohoku Denryoku officially announced their interest in Maki. The group began sending inquiries to the company and protesting at the town councils.

When Maki town council approved the construction of the first reactor in 1977, the antinuclear movement fired up. Despite the upsetting turn of events, several internal and external conditions were in the opposition's favor. The land issue was one. Years before, Endo bought from a Gokahama resident a small piece of land (about 1,800 sf.), which sits in the middle of the proposed site. According to Mrs. Kuwabara, his motive then was not about using the land as a resistance leverage. Rather, he bought it so that he would not abandon his activism. Later, this land proved to be invaluable in Maki's antinuclear movement. The proposed site was a patchwork of small parcels of land owned by individuals and the town. In order to build a plant, all these parcels had to be procured. Anticipating this challenge, Tohoku Denryoku founded a separate company before the 1969 exposé to begin purchasing land under the pretense of tourism development (Ito et al., 2005). Still, not all landowners agreed to sell, particularly after the exposé came out. Six others joined Endo as co-owners of the small piece of land and formed *Genpatsu hantai kyoyu jinushi kai* (an antinuclear landowners' group) to challenge the town council's pronuclear decision. Additionally, a lawsuit was underway over the ownership of a cemetery that stood where the first reactor was to be built. These and other pieces of land that *Tohoku Denryoku* was yet unable to acquire forced the company to give up the construction of additional reactors and to suspend the construction of the first reactor for ten years (Kuwabara & Kuwabara, 2003).

There were other factors in the 1980s that contributed to the suspension. As the oil price was low in the 1980s, utilities found importing thermal power cheaper than constructing nuclear power plants (Ito et al., 2005). Politically, Maki's dominant conservative camp was divided into two, and the one that captured vacillating voters always won mayoral elections (Kuwabara & Kuwabara, 2003). Since those voters included people who were not in favor of nuclear power coming to their town, the mayoral candidates had to take a "cautiously pronuclear" stance and was not able to aggressively support the nuclear program. Then, the Chernobyl nuclear disaster occurred, prompting strong and widespread antinuclear reactions throughout the country. In Maki, too, even previously silent residents voiced their opposition to nuclear power.

Citizen-centered movement

Although all aforenoted external and local conditions were indubitably important, they alone would not have been sufficient for the opposition to prevail – not when the town council, the mayor (albeit ostensibly "cautiously pronuclear"), the governor, and the fishery association were all in favor of the nuclear project. The opposition would not have stood a chance without the local activists who leveraged the conditions and remained committed to their cause. Maki's antinuclear movement eventually led to a referendum that voted down Tohoku Denryoku's nuclear project in Maki. What made this possible? The activists believe that it was because the movement was citizen-centered. The movement started this way when Endo and his friends engaged the residents of Gokahama. However, the landowner's group strayed from this beginning until they realized its importance. Mrs. Kuwabara shared the following story:

> Mr. Endo's antinuclear landowners' group was not focused so much on communicating with the residents. Then, one day that changed. Tohoku Denryoku did an environmental survey in 1981 and announced that they would be holding information sessions to share the survey results. Endo's group picketed to protest the sessions. In response, Tohoku Denryoku pressured the business bureau and mobilized local business owners to picket against Endo's group. There, Endo and others in the group saw on the adversary side their friends and those men who were like uncles to them. They realized that these opposing picket lines meant a nasty shouting match against the people dear to them. They realized that everyone became a puppet of Tohoku Denryoku. By fighting against each other, no one wins but Tohoku Denryoku. They realized that their movement had strayed but now needed to get back to engaging their fellow residents and not fighting them.

This return to engaging the citizens then led to the formation of a group of antinuclear activists who supported Tamio Takashima in the 1982 mayoral election. Seeing that there was no candidate sympathetic to the antinuclear cause, Takashima, an antinuclear lawyer, declared his candidacy only two months before the election. A group of 20–30 residents came together to support his campaign under the motto of keeping Maki free of a nuclear facility. Among the activists were Mr. and Mrs. Kuwabara who were both teachers and had moved to Maki, Mr. Kuwabara's hometown, a few years ago. The couple, along with another teacher, launched a campaign calling for Maki citizens to voice their opinions against nuclear power. Then, they joined with others to form a group to support Takayama's candidacy. Mrs. Kuwabara recalls this experience:

> This body was independent of any existing political organizations. As we worked on the election, we built trust, shared common objectives, and fostered perseverance. We were determined to do our best even knowing that we would lose. These are the ingredients of a citizen movement, and we nurtured them as we worked together for the election. It was a very intense experience. Without that, I don't think the citizen movement would have been born in Maki.

Although Takashima lost the election by a large margin, it was significant that some 2,300 people (12%) voted for him. For the opposition, this signaled that many Maki residents do not want a nuclear facility to their town.

The day after the election, the group renewed their vow to continue fighting to keep Maki nuclear-free. They engaged the residents in a variety of ways. They held seminars for the residents to learn and discuss topics related to nuclear power. They organized tours and hikes for the residents to show the natural beauty of the coast and adjacent mountains that would be destroyed if a nuclear power plant were to be built. For ten years, they published monthly newsletters and inserted them in the local newspaper to be distributed to every household. Separately, Mrs. Kuwabara and two other women in the group – all teachers – wrote a monthly newsletter focusing on women (e.g., potential danger of nuclear power to women and children and women's voices in the antinuclear activities around the country) and distributed in front of the train station for six years. According to Mrs. Kuwabara, all through these activities, they stayed true to the principle of a citizen-centered movement. The group believed that, if they have any chance of success in nullifying the construction plan, they must continue to engage the residents by creating spaces for discussing potential dangers of bringing a nuclear power plant, even while the project was dormant. Endo believed that it was this unwavering questioning of nuclear power that eventually led to the referendum (Ito et al., 2005).

I will return to the referendum shortly, but there was another aspect of Maki's antinuclear movement that deserves a mention. Its commitment remained resident-centered not only in its engagement but also in the composition of the movement itself. When asked about the relationship between Maki's movement and other antinuclear movements, Mrs. Kuwabara emphasized that there was very little relationship, because their movement was by and for the local residents:

> We were convinced that Maki's nuclear problem must be tackled by us, the residents, not by outsiders, because our movement relied on visiting each home and talking to the residents. We were appreciative of the help outsiders wanted to extend to us, but we wanted to do this on our own. We were even called "Maki Monroe Doctrine." We accepted volunteers from other parts of Niigata, but otherwise we didn't want outsiders who were unfamiliar with our town to come in and criticize pronuclear residents. They wouldn't understand what may be driving people to support nuclear power.

The last comment reflects the complexity of the politically conservative and patriarchal culture that dominated Maki at that time. Residents might have supported the nuclear project because they are pressured to go along with the city and business leaders who were predominantly pronuclear. This is a nuance that outsiders might not have understood and could have harmed the opposition's effort – a risk that the Maki activists did not want to take.

Empowering silenced voices: cranes, handkerchiefs, and referendum

Although the opposition continued to engage the residents, the nuclear project again surfaced as a key agenda during the 1994 mayoral election. By then, the fear of radiation poisoning that filled Japan's air after Chernobyl dissipated, and the pronuclear voices were once again louder. The incumbent Maki mayor, Sato, declared that, upon his third victory, he would make Maki home to the world largest nuclear power plant. Because Sato was a powerful incumbent, the activists felt that his reelection could mean defeat. Yet they could not simply abandon the movement without putting up the last fight. They knew that there were mothers who opposed nuclear power for the sake of their children but could not voice it because women, especially daughters-in-law, were supposed to stay out of politics. Mrs. Kuwabara explained that Maki in the 1980s was still firmly gender-stratified; it was understood that women's place was in the kitchen and nowhere else. She wanted to find a way for women to express themselves without outwardly challenging patriarchy. After much contemplation, she came up with the idea of paper cranes. She thought that women could make paper cranes to show their opposition to the nuclear project. She and a few others made a flyer about this idea and distributed it. To her surprise, it ignited a fire across the town, and within four months, they collected 130,000 cranes.

It was with this backdrop that the 1994 mayoral election took place. Besides the incumbent pronuclear Sato, there were two other candidates:

one was antinuclear and the other called for a town-wide opinion survey regarding nuclear power. Although Sato won, the other two candidates combined received more votes. This meant that the majority of the residents either opposed the nuclear project or at least wanted to know how everyone felt about the project. Realizing this, the opposition, made up of separate groups, formed a coalition and demanded the city to hold a referendum. Unsurprisingly the mayor opposed the idea, but that did not deter the opposition. Instead, they held a grassroots referendum. It took them three weeks to complete it, because they made sure that everyone who wanted to vote had an opportunity to do so. According to Mrs. Kuwabara, the city government and city's powerful voices such as business associations were all in favor of constructing the Maki Nuclear Power Plant and pressured residents not to vote. But the residents did. There were young fathers who came right before the closing time to quickly vote before anyone saw them. There were residents who came to vote, wearing a hat and a mask to hide their faces. There were grandmothers who walked on rough roads to vote. Despite the pronuclear intimidations, 45% of the eligible voters participated in this unofficial referendum, and the result was overwhelming; 95% of the voters opposed the nuclear project.

With this result as the warrant, the opposition pressured the mayor to hold an official referendum. He again refused but later resigned when the opposition threatened to impeach him. His resignation was followed by a new mayoral election, which was won by an antinuclear candidate, Takaaki Sasaguchi. This victory made it possible for the first official referendum in Japan to take place a year later in 1996. Leading up to the referendum, the opposition again engaged their fellow residents in interpersonal conversations. Mrs. Kuwabara stressed the importance of keeping the movement personal and local:

> We wanted the residents to understand that the nuclear project we are facing is not about energy but about Maki's daily life. My husband used to say that, if there is a clogged ditch in front of your house, you want to unclog it. We wanted the people of Maki to think of the nuclear project as something like that. It is not a political issue or an agenda of a political party. It is our everyday life issue.

For this reason, the movement remained local in the makeup of the activists and their approach. In addition to home visits, the activists also solicited residents to write a message on a handkerchief. The handkerchiefs read, for example: "I don't want wealth if it means to live with danger"; "life or money, which is more important? (6th grade student)"; "stop nuclear power for the sake of children"; "26 years of lies. No more" (Kuwabara, 2014). Trees full of these handkerchief messages were created and displayed around the town to show that a large number of residents are against the construction of a nuclear facility.

The official referendum attracted impressive 88% of the eligible voters and passed with 60% of the votes opposing the construction of a nuclear power plant. Between this result and the antinuclear mayor in the office, it may appear that Maki's antinuclear movement was winning. However, as 13 of the 22 members of the town council were pronuclear, the nuclear project was still not off the table. Tohoku Denryoku announced postponement of the project for three years, but they maintained their intention to pursue it. To prevent the project from resurfacing, Mayor Sasaguchi sold a piece of town's land within the planned construction site to the opposition. This move infuriated pronuclear council members who subsequently sued the mayor, but the Niigata District Court sided with the mayor, concluding that the mayor's action was lawful and was consistent with the citizens' wish expressed through the referendum. The plaintiffs appealed to the Tokyo High Court and then to the Supreme Court, but the appeals were dismissed. With the dismissal, the president of Tohoku Denryoku announced in 2004 that the company would no longer pursue the Maki nuclear project.

Ingredients of successful antinuclear activism at ground zero

The antinuclear movements in Hidaka-cho and Maki-machi both lasted over 30 years. It should not be forgotten that the communities, even if they successfully resisted nuclear power, suffered enormous relational, psychological, and financial costs as a result of the site fight that spanned three decades. All that time and energy could have been used to build economic and social resilience unique to the community (Hirabayashi, 2013). With that understood, they are remarkable displays of grassroots resistance to the nuclear industrial complex that had vast financial and political resources at their disposal. Each community had their unique local circumstances and fought their battle with the hands that were dealt. Still, the stories reveal some common attributes that helped the two movements to prevail. Given that the movements were occurring around the same time period, there were external, historical conditions that worked in their favor. The news of the accidents at the TMI and Chernobyl raised awareness about the danger of nuclear power. The drop in the oil price shifted the power companies' attention away from nuclear power temporarily. But these conditions were supplemental. The keys to the movements' victory were something else.

In both communities, the opposition owned something that the power company absolutely needed to win the site fight. In Oura, this key piece was the fishery association's consent, and, in Maki-machi, it was small pieces of land within the proposed construction site. The consent of the local fishery association is one of the first essential hurdles that must be cleared for a nuclear facility construction project to proceed. If the association gives in to financial incentives, which are always substantial, resistance becomes immensely difficult. Similarly, landowners refusing to sell a piece of land in the middle of a proposed site can break a deal. As a nuclear power facility requires a vast site that meets a number of geological, geographical, and maritime criteria, it is entirely critical that a project owner secures the site they selected. This is why, as we saw in the case of Hidaka-cho, power companies start buying lands under disguise years before they officially announce their plan to build a nuclear power plant.

Withholding consent and land are essential, but they alone do not guarantee a victory. There were two attributes that the two movements had in common. One was perseverance of devoted activists. In both towns, the activists demonstrated extraordinary determination to continue their fights even in the face of multiple losses, pains, and betrayals. Another attribute was their faith in and love for their community. In Hidaka-cho, we saw the activists appealing to the comradery of the fishing community and their common obligation to protect the community for future generations. In Maki-machi, the activists tirelessly cultivated community relationships through group activities and one-on-one visits with fellow residents.

The desire to maintain the integrity and relationship within a community is not unique to Hidaka-cho and Maki-machi. Anthropologist Kouhei Inose (2016) studied a nuclear power site struggle in Kubokawa, Kochi Prefecture, and concluded that the common view that a nuclear power town creates polarization between pronuclear and antinuclear is misguided; the residents in Kubokawa had the wisdom to work out conflicts thanks to the deep-rooted relationships they had cultivated for generations through sharing agricultural work and social life. Kubokawa's site fight began at the dawn of 1980 and lasted through the decade. The town made a national news when they proposed to use of a referendum over the construction of a nuclear faculty. As discussed in Chapter 3, *The Asahi Shimbun* viewed the referendum as a good experiment of how to best reflect local communities' wishes ("Shasetsu: Jūmintōhyō," 1982, June 30), whereas *The Yomiuri Shimbun* saw it as a selfish action that undermines the needs of the larger public and the authority of a representative democracy ("Shasetsu: Genpatsu," 1982, July 1). From an insider's perspective, Inose wrote that the referendum was the last resort because it could leave a deep scar in the community. The residents knew that and diligently held town meetings to learn more about nuclear power and discussed the implications of hosting a nuclear power plant. In the end, they rejected the idea without resorting to a referendum. Inose underscores that Kubokawa was able to do so because the residents understood their responsibility to protect their collective well-being and engaged in dialogues.

The cases of Maki-machi and Hidaka-cho, along with Kubokawa, teach us the significance of civic society. Their examples give support to Aldrich's (2008, p. 15) argument that local civic societies "support *or* contest state initiatives, making it less or more difficult for states to enact policies." Thus, after geological and geographic criteria, the strength of civic society is a critical consideration for site developers in making decisions about the best sites for controversial facilities like nuclear power plants. They examine the quality – "the strength and depth of the bonds between citizens" and relative capacity (the size of the group in comparison to their opposing groups) of civic society at potential sites (Aldrich, 2008, p. 31). Communities with a strong and widely networked civic society are in a much better place to resist the state-industry social control strategies and win site fights. And that was what Maki-machi and Hidaka-cho proved through their three decades of activism. These communities also exposed the flaw in the logic of colonization (Plumwood, 1993) where the nuclear industrial complex assumes the master position or superiority that incorporates and instrumentalizes local communities. The movements showed that communities have the power to refuse being reduced to an instrument for the master and being defined by the master's monetized value. Hence, they serve as examples of disruptions to the dominant institutions.

Howard Zinn (2007) argued that articulating hidden stories of resistance is important because they can show new possibilities for the future. The local antinuclear power movements reveal the power citizens have, but much of that history remains hidden or overlooked even inside Japan. This absence from the public consciousness contributes to emotional remoteness (Plumwood, 2002) to antinuclear struggles, maintaining the pronuclear hegemony. As discourse rules in as well as rules out meanings (Hall, 1997), it creates or undermines possibilities for the lifeworld. Without knowing these stories, it is easy to accept when the largest newspaper writes that an important national policy matter like nuclear power should not be decided by local referendums. It sounds entirely logical. The post-Fukushima public outcries, which will be discussed in Chapter 7, give some clue that different citizenry with regard to nuclear power might have been possible if these stories of activism to fight for their communities were widely publicized. Even before the Fukushima disaster, however, there were citizens who devoted to telling stories to challenge pronuclear discourse, a topic to which I now turn.

Notes

1 Beginning in 1962, a lumber company from Osaka approached the Ao district of Hidaka-cho, stating that they want to purchase a land to build a sawmill (Asami, 2012). The company president promised to the landowners that he would never sell the land, and, if he is unable to build a sawmill for any reason, he would sell back the land (Terai, 2012). With this promise, the landowners collectively agreed to the sale. However, no factory went up. Six years after the sale, the land was sold to KEPCO. Infuriated, 140 landowners sued the lumber company in 1972, demanding the company to return the land to them (Terai, 2012). However, in the 1979 decision, the Wakayama prefectural court sided with the company, stating that the promise was not binding and denied that the company committed a fraud. The company then sued the Ao residents for damages, which was later dropped.

2 These nuclear scientists included Touru Ebizawa, Shinji Kawano, Hiroaki Koide, Keiji Kobayashi, and Ken Seo. None of them were promoted and remained

assistant professors or lecturers despite the decades of employment, while their colleagues who support nuclear power were promoted to full professorship. For their courage and conviction, they are profoundly respected among Japan's antinuclear activists.

3 About US$1.3 million at the 1984 exchange rate.
4 About US$5.36 million at the 1988 exchange rate.
5 About US$17 million at the 1980 rate.

References

Aldrich, D. P. (2008). *Site fights: Divisive facilities and civic society in Japan and the West*. Ithaca, NY: Cornell University Press.

Asami, K. (2012). Wakayama ni genpatsu ga yatte kuru [Nuclear power is coming to Wakayama]. In *Genpatsu wo kobamitsuzuketa Wakayama no kiroku [The history of the antinuclear activism in Wakayama]* (pp. 20–40). Sapporo, Japan: Jyuroushya.

Behling, N. H., Behling, M. C., Williams, M. C., & Managi, S. (2019). *Japan's quest for nuclear energy and the price it has paid: Accidents, consequences, and lessons learned for the global nuclear industry*. Cambridge, MA: Elsevier.

Fisher, W. R. (1984). Narration as a human communication paradigm: The case of public moral argument. *Communication Monograph, 51*, 1–22. https://doi-org.libproxy.uncg.edu/10.1080/03637758409390180

Fisher, W. R. (1985). *Human communication as narration: Toward a philosophy of reason, value, and action*. Columbia, SC: University of South Carolina.

Hall, S. (1997). The work of representation. In S. Hall (Ed.), *Representation: Cultural representations and signifying practices* (pp. 15–64). Thousand Oaks, CA: Sage Publications, Inc.

Hirabayashi, Y. (2013). "Genpatsu okotowari" chiten to hangenpatsu undo ["No nuclear power" sites and antinuclear power movement]. *Ohara shakai mondai kenkyujo zasshi, 661*, 36–51. https://doi.org/10.15002/00009519.

Inose, K. (2016). *Mura to genpatsu: Kubokawa genpatsu keikaku wo momikeshita shimanto no hitobito [Village and nuclear power: The people of Shimanto who stamped out the Kubokawa Nuclear Power plan]*. Tokyo: Nobunkyo.

Ito, M., Watanabe, N., Matsui, K., & Sugihara, N. (2005). *Democracy Reflection: Maki Jyumin touhyou no shakaigaku [Sociology of Maki referendum]*. Tokyo: Liberuta Shuppan.

Kuwabara, M. (2014). *Genpatsu nanka iranai, anotoki Maki-machi de okitakoto [No nuke: A chronicle of the challenges that Maki-machi overcame]*. Presented at Movements Oneness Meeting, December 29, 2014, Tokyo, Japan. Retrieved from http://onenesscamp.org/docs/NoMoreNuke_TheThingsHappenedInTsuruMaki-Cho_AtTheTime.pdf

Kuwabara, M., & Kuwabara, M. (2003). *Maki Genpatsu, Jyumin Tohyo eno kiseki [Maki Nuclear Power Plant, the miracle that led to the referendum]*. Tokyo: Nanatsumori Shokan.

Nakanishi, H. (2012). Hantai undou wo dou tatakatte kitaka [How the opposition movement was fought]. In *Genpatsu wo kobamitsuzuketa Wakayama no kiroku [The history of the antinuclear activism in Wakayama]* (pp. 188–213). Sapporo, Japan: Jyurōshya.

Plumwood, V. (1993). *Feminism and the mastery of nature*. London: Routledge.

Plumwood, V. (2002). *Environmental culture: The ecological crisis of reason*. London: Routledge.

Shasetsu: Genpatsu no jūmintōhyō wo kangaeru [Editorial: Thoughts on Nuclear Power Referendums]. (1982, July 1). *The Yomiuri Shimbun*, p. 3.

Shasetsu: Jūmintōhyō no jikken wo mitsumeru [Editorial: Observations on the Referendum Experiment]. (1982, June 30). *The Asahi Shimbun*, p. 5.

Squire, C. (2013). From experience centred to socioculturally-oriented approaches to narrative. In M. Andrews, C. Squire, & M. Tamboukou (Eds.), *Doing narrative research* (pp. 47–71). London: SAGE Publications, Inc.

Terai, T. (2012). Gokasho no genpatsu keikaku to hantai undou [Five nuclear project plans and antinuclear movement]. In *Genpatsu wo kobamitsuzuketa Wakayama no kiroku [The history of the antinuclear activism in Wakayama]* (pp. 42–141). Sapporo, Japan: Jyurōshya.

Yamaaki, S. (2012). *Genpatsu o tsukurasenai hitobito [People who fight the construction of a nuclear power station]*. Tokyo: Iwanami Shoten.

Zinn, H. (2007). *A power governments cannot suppress*. San Francisco: City Lights Books.

6 Pre-Fukushima urban antinuclear activism
Identity and sociocultural challenges

Pre-Fukushima urban antinuclear activism: why important?

Chapter 5 took a close look at the antinuclear movements in local communities where fights over the construction of nuclear power plants were intense, arduous, and long-lasting. Antinuclear movements existed in urban Japan as well long before the Fukushima nuclear disaster. It is worthwhile to shed light on these urban movements separately from the local ones, as they had a different role to play in the overall antinuclear movement. If local communities were the ground zero of blocking each project, urban movements tackled the nuclear power program as a matter of national policy that citizens must criticize and reject. Toward this end, urban activism took a variety of forms: serving as a hub for information exchange and networking, building a sense of larger community between local communities and between local communities and cities; educating urban residents about the danger of nuclear power and the struggles of the local communities through seminars and study sessions; visiting utilities' headquarters and government offices to demand transparency, exposing the misconduct of utilities and national and local governments; filing lawsuits; and holding rallies. All these roles were played by the urban activists I interviewed.

This chapter features the experiences and perspectives of veteran activists from three metropolises – Tokyo, Osaka, and Kyoto. As my interest lies with the activists' perspectives, I again follow a narrative approach to feature the stories of pre-Fukushima urban antinuclear movements through the lenses of the veteran activists. Originally, my goal was to chronicle major moments of the movements from the points of view of the activists. However, the inadequacy of this approach quickly became clear. Unlike local site fights where the antinuclear movements have a clear, single goal of stopping the construction of nuclear power facilities in their communities, urban movements addressed the overall nuclear power program and were ongoing and often unheralded. Thus, my conversations with the activists centered on their experiences of voicing against the program in pre-Fukushima Japan and the obstacles they encountered along the way. In the following, I first briefly consider the beginning of the urban antinuclear movement that focused on nuclear power issues. The rest of

DOI: 10.4324/9781003044222-6

the chapter is devoted to the seven veteran activists I interviewed. I discuss their motives for activism and the themes that emerged from the interviews, including activism as identity, passive democracy as a barrier to social activism, social marginalization of antinuclear power movements, and media as a barrier to antinuclear power movements.

The birth of urban antinuclear power movements

It is possible to see the 1960s as the beginning of the urban movements that exclusively addressed nuclear power. As soon as the plans to build nuclear power reactors in various parts of Japan emerged in the 1960s, local opposition arose, and national organizations with headquarters in Tokyo such as *Sohyo* (the General Council of Trade Unions of Japan that was disbanded in 1978), *Gensuikin* (Japan National Conference Against Atomic and Hydrogen Bomb), and the Socialist Party supported the local resistance (Honda, 2019). Although these oppositions from the national organizations are important to recognize, they did not necessarily coalesce into a coordinated, large-scope social movement (Stewart, Smith, & Denton, 2012) centered on nuclear power. Additionally, some of these organizations are institutions themselves albeit nonmainstream ones and thus did not represent an uninstitutionalized collectivity (Stewart et al., 2012) characteristic of social movements.

A more intentional, collective, enduring urban organizing against nuclear power emerged in the mid-1970s. From August 24 to 26, 1975, citizens and groups around the country who oppose nuclear power gathered in Kyoto for the first time. The conference was convened to object the government-sponsored public hearings about the construction of the Kashiwazaki-Kariwa Nuclear Power Plant in Niigata Prefecture (Nishio, 2013). The government implemented double public hearings: a central symposium in Tokyo (far from the site) where experts discussed safety-related matters and a public hearing in Kashiwazaki-Kariwa that focused on the economic benefits of hosting a nuclear power plant. The activists who gathered at the conference objected to the deliberate design of public hearings that were held after-the-fact (after the land was all procured) and excluded the Kashiwazaki-Kariwa residents from voicing their concerns about nuclear power plant safety (Nishio, 2013).

One of the challenges that the conference participants expressed was the lack of means to share information between regions (Sueda, 2018). In the following month, nuclear scientist Jinzaburo Takagi founded the CNIC, a public interest antinuclear organization in Tokyo. Takagi gave up a prestigious, lucrative career in the nuclear industry to establish the CNIC to dedicate his life to providing reliable information about nuclear power and public education in order to ultimately secure a nuclear-free world (CNIC, n.d.). In response to the growing need for mediums to connect regions and share local news, the CNIC led the establishment of *Hangenpatsu Undo Zenkoku*

Renraku kai (the National Liaison Conference of Antinuclear Movements) in 1978. This national liaison, housed in the CNIC, then began to publish *Hangenpatsu Shinbun* (the antinuclear power newspaper), a monthly digest of antinuclear activities and related information around the country with the purpose of learning from each community's fight and of transforming the separate fights into a united movement (Sueda, 2018). The newspaper has published over 500 issues to date.

Since the mid-1970s, urban antinuclear power movements rather quietly dwelled in metropolises.[1] Sociologist Takemasa Ando (2019) sees urban antinuclear movements as part of what is called *orutanatibu undo* (the alternative movements) such as fair trade, coops, recycling, and organic food that emerged as a reflexive response to the rapid economic growth that Japan saw in the 1960s. Nuclear power, then, was seen largely from a public health perspective. It is fair to say that it was not until the late 1980s – after the Chernobyl nuclear accident of 1986 – that the movements became a more visibly active and sustaining presence. As the concerns over contamination of food from the Chernobyl radiation traveling through the air rose, housewives in Tokyo, for example, began to organize against nuclear power (Ando, 2019). However, Chernobyl was only one of catalysts for my interviewees. As the next section shows, there were other reasons that drove them to activism long before Chernobyl.

Becoming an antinuclear power activist

Baku Nishio's history of antinuclear power activism can be traced to early 1970s when he worked for an advertising agency as a young adult. At that time, there were local oppositions to the construction of coal-fired power plants, which frustrated electric utilities. To control the oppositions, the Federal Electric Power Companies of Japan (FEPC) launched a nationwide advertising campaign (large monthly advertisements in newspapers and magazines) in 1973 to instill fear in consumers that there will be an electricity shortage crisis if more plants are not built. According to Nishio, these ads were aimed at urban consumers and were designed to encourage them to see the local resistance to the plant construction as their enemy. Not only did advertising companies work with FEPC to produce these ads, they also became FEPC's extension; they researched the weaknesses of the residents of the communities that were resisting coal-fired plants in order to blackmail the residents to change their minds. These manipulations shocked Nishio so profoundly that he started an anti-advertisement group to criticize the role of advertising in perpetuating the oppression. Soon, utilities began to do the same with nuclear power plants, which prompted him to learn more about nuclear power. A few years later in 1978, he joined the CNIC to help launch the *Hangenpatsu Shinbun* project. Later, he became one of the three codirectors of the CNIC.

Makoto Yanagida's entry into activism was through a pollution and public health study group he joined in early 1980s while he worked for the Tokyo government. The group gathered to learn about severe cases of illnesses such as Yokkaichi asthma, Minamata disease, and the illnesses caused by the pollution at the Ashio Copper Mine. When the Chernobyl accident occurred, the fear of radiation poisoning was widespread in Japan. His study group wanted to measure the radiation levels in food, and his well-off friend provided the funds to purchase an expensive radioactivity measurement device. This became a turning point for the group to formalize its activity; the members rented a space and established Tampoposya in 1989. Yanagida felt that nuclear power is the biggest threat to public health because it could cause a catastrophic accident. Tampoposya has been a nuclear power watchdog, publishing information about nuclear power, holding seminars, publishing newsletters, and demonstrating on the streets of Tokyo calling for the abolishment of nuclear power.

One of the most energetic antinuclear activists in Japan is Aileen Miyoko Smith. Smith has a long history of activism since the late 1960s when she got involved in the anti–Vietnam War movement at Stanford University where she was studying. After learning more about injustices around the world, she left the university because she felt that being part of social justice movements was urgent. In 1970 she met her first husband, Eugene Smith, a photojournalist, in New York. They both learned about Minamata disease, a neurological disease caused by mercury poisoning from industrial wastewater dumped to Minamata Bay by a chemical company, Chisso, between 1932 and 1968. They both spent a few years in Japan, photographing the chronic fights of the Minamata victims for justice. Then, there was a mercury poisoning outbreak among the Ojibwe people in Ontario, so Aileen helped to organize the exchange between the two communities. With her second husband, a scientist and antinuclear activist, she also visited the communities around the TMI nuclear power plant after the 1979 nuclear accident to hear and record the experiences of the residents whose voices were ignored by the U.S. government and journalists alike. After spending a year there, they returned to Japan. They were invited to speak at towns across Japan where nuclear power site fights were taking place, and she became increasingly involved in antinuclear power activities. In 1991, she attended an international plutonium conference in Tokyo organized by the CNIC. At the conference, she learned about the danger of the plutonium recycling program and felt the need to create a network and resource space to address the danger. That led to the establishment of an NGO, Green Action, in 1991 with the aim of creating nuclear-free Japan. As a Japanese-English bilingual, Smith publishes documents in both languages and often speaks about Japan's nuclear power developments to the international community.

Hideyuki Koyama is the head of Mihama-no-kai, an antinuclear power organization in Osaka. Like Smith, Koyama has a long history of social activism that began in college when he participated in the mass demonstrations against the U.S.-Japan Security Treaty of 1960 (known as *Anpo*). He was also active in the national graduate student conference that advocated for improving graduate students' lives. After graduation, he joined the labor movement. His initiation into antinuclear activities was the TMI accident. Previously he vaguely thought that nuclear power was dangerous, but TMI was the first time he witnessed the danger through the news. Prompted by the accident, he started a nuclear power study group with students and fellow lecturers at the university where he lectured. He also began to travel to the communities where site fights were taking place to assist to the local movements such as the one at Hidaka-cho that was featured in Chapter 3. After the No. 2 reactor at the Mihama Nuclear Power Plant in Fukui Prefecture experienced an accident and released radiation in 1991, Koyama founded Mihama-no-kai to file formal complaints about the danger of nuclear power plants.

Eri Fujikawa's involvement with antinuclear movements began after the Chernobyl accident that occurred when she was a college senior.[2] The accident shook her, and she began to read about nuclear power. Previously she thought that nuclear technology was necessary and that small-scale problems and accidents were a small price to pay. Chernobyl and her subsequent study about nuclear power changed her mind. Still, her main areas of activism remained gender and ethnic civil rights issues as well as antiwar. The first real wake-up call was the fire at the Monju Nuclear Power Plant in 1995, as it was close to Kyoto where she lived. If there was a serious accident at Monju, it can annihilate Kyoto, she thought. Then, the Tokai-mura nuclear accident that killed two of the three workers occurred in 1999. Fujikawa learned that the workers died in a way very similar to the firefighters who were killed at Chernobyl: "The firefighters were sent to the scene without the knowledge of how dangerous it was and were exposed to the deadly level of radiation. Their muscles melted, and they bled to death. And that is how the workers at the Tokai-mura plant died, too." She began to participate more actively in study sessions, talks, and gatherings on nuclear power. In 2003, she launched a magazine to address a variety of social justice issues, including nuclear power, with the aim of contributing to the creation of a peaceful, just Japan.

Hiro Murata became interested in nuclear power issues in 2004 when he watched a documentary film that revealed the sicknesses, birth defects, and deaths among the Iraqi children due to the exposure to depleted uranium ammunition during the Gulf War.[3] He learned that these weapons were made with the wastes produced in the process of producing nuclear power fuels and that the U.S. company that produced the fuels (thus produced the depleted uranium) also produced fuels for Japan's nuclear reactors. He was shocked by the film and started to host community screenings of the film so more people learn about how Japan's nuclear power is connected to the radiation victims in Iraq. Since then, he started organizing community film screening and dialogues about nuclear power and began attending nuclear power study groups and meetings. He became one of the main organizers of the annual gathering and music concert against nuclear power in Ōma,

Aomori Prefecture. The annual event began in 2008 when the construction of the Ōma Nuclear Power Plant began. This plant is unique in that it is the only nuclear power plant in the world that exclusively uses plutonium-uranium MOX fuels – a mix of depleted uranium and reprocessed plutonium. Antinuclear activists are particularly concerned about Ōma because MOX fuels are suspected to be more prone to accidents.

Finally, Junichi Sato served as Director of Greenpeace Japan until 2016. The seeds of his activism were planted when he studied abroad in the United States. One of his classes took students to a demonstration against nuclear experiments in Nevada, and he was able to hear both sides of the issue. Then, as an anthropology major, he spent a year in Mexico with the Rarámuri people and learned about the social and economic oppression and environmental destruction that this indigenous group was facing. These experiences led him to seek a career in an environmental nonprofit. In a small book about activism published by Greenpeace (2012), Sato further explained that his particular interest in Greenpeace arose after attending a nonviolent direct action workshop given by a Greenpeace volunteer in Nevada. He was impressed by this approach to activism that was based on detailed planning and strategic communication with their opponents rather than reacting haphazardly without a plan. He joined Greenpeace Japan in 2001.

Antinuclear power activism as identity

One apparent common thread among the urban activists I interviewed is the existence of antecedents to their antinuclear power activism. Except for Murata, everyone was already engaged in some forms of activism whether it was labor rights, gender discrimination, antiwar, or pollution. In western discourses, both academic and nonacademic, modern Japan has often been brought up as an example of a collectivist culture that values harmony and group goals.[4] But this is an oversimplified and even erroneous representation that fails to account for the oppositional and dissenting voices that existed in the post-WWII Japan. The massive demonstrations against the passing of the revised security treaty (known as *Anpo*) that Koyama referenced, for example, represent the fierce, widespread public objections to the government that undermined peace and democracy in Japan. The Korean War brought a war boom to Japan, but this was mentally, morally, physically taxing. Many workers were forced to work long hours at meager wages to produce bombs, ammunition, and other supplies for the United States in the war against their neighboring Korea (Yamamoto, 2004). This and the dominant presence of U.S. military bases in Japan catalyzed labor unions, students, and women's organizations, among others to stage multiple strikes to prevent the conservative government from making any revisions to subject Japan to a further U.S. control (Jesty, 2012). In the 1960s, the largest of the *Anpo* strikes drew 6.2 million workers (Packard, 1966/2015). Thus,

Japan's metropolises saw robust social and labor movements from the 1950s to the 1970s.

Koyama came into adulthood in the backdrop of these antiwar and democracy movements. Participating in social movements was part of his socialization. Fujikawa, although a generation younger than Koyama, also developed a sensibility for social movements as she grew up hearing her parents' stories about the atrocity of the war they experienced and the importance of protecting Article 9 of the Japanese Constitution that prohibits wars. As a result, it was natural for her to engage in antiwar and antidiscrimination activism. Smith was already an antiwar activist in college. Yanagida was actively addressing pollutions and so was Smith. And Sato's activism was sparked during his college years, and he was already active in addressing human rights and environmental issues through Greenpeace.

The paths that led the urban antinuclear power activists to oppose nuclear power contrast with those of the local antinuclear activists. Whereas the local activists were largely thrown into the antinuclear power movement because nuclear facilities were coming to their own communities, the urban activists were already social justice activists before getting involved in the antinuclear movement. For them, their antinuclear activism was an extension of these previous experiences. Fujikawa articulated this link:

> During the war, my father barely avoided the major air raid in Osaka. I grew up hearing horror stories of wars from my parents. So, I absolutely oppose wars. Nuclear power is no different. We already know that it will harm health, then why use it? Using it is no different from engaging in a war. So, for me, wars, Article 9, nuclear power, they are all connected.

For the veteran activists, participating in antinuclear power movements was part of their identity, much like the long-term activists in the anti-TMI movement who felt that their activism was already in their DNAs (Angelique & Culley, 2014). Smith explained this idea of activism as identity in the following way:

> I just can't stand watching injustice. It just makes me mad. It just makes me really perturbed. Angry. It's sort of impossible to be silent. But also it's really an interesting job using your brain. It's like David and Goliath. You know, how to be much smaller and . . . I always say it's like judo and shiatsu where the pressure point is. Or use the power on the other side to throw them. Like aikido. It's actually very exciting and interesting work.

Smith's reference to David and Goliath and aikido (a Japanese marshal art) represents antinuclear power activists as righteous underdogs with a sense of moral obligation to act for the good of people (Stewart et al., 2012). Being an activist in this sense is an outward expression of their core value.

It is also noteworthy that, despite the serious and overwhelming nature of the problems that she is up against, Smith's view of activism as identity is not a stressful one. She finds it "exciting and interesting." I saw similar expressions of ease and perseverance in other activists I interviewed. When asked what drives him to persevere, Koyama laughed and gave me the following answer:

> Well, I don't really think about that. It's like a force of habit. It's my life. This [how to stop nuclear power] is something I think about from the morning until I go to bed. I even think about it in bed. It's a force of habit. I think it's the same for others.

Murata agrees that it is part of his life. However, unlike Smith and Koyama who lead antinuclear power organizations, Murata has his own family business. His response reflected this background.

> For me, it is part of my daily life. I don't really think about it as "anti." It's just that the problem of nuclear power is so morally wrong that I want many people to know about it. . . . But I don't get too zealous about it. I want to enjoy life, and I made it part of my lifestyle. For example, if I want to buy a bicycle, I don't look for the best deal. I try to see if there is a bike store that is owned by someone who oppose nuclear power. I want to support their business. . . . I'm really busy but, every week, I go to a boxing gym and a tap dance class. Occasionally, these classes conflict with meetings on nuclear power issues, but I stick to the classes. My activism is part of my life, and, to continue it, I need to be able to do other things, too.

In sum, all the veteran activists I interviewed were driven by a sense of obligation to fight what they perceive as the moral bankruptcy of the nuclear power industrial complex. However, each activist has carved out their own ways to make the activism their identity and everyday life.

Passive democracy as a barrier to social activism

While activism was part of their identity, and they were not afraid to voice their opposition to what they saw as injustice, the urban activists experienced considerable difficulties and frustrations. Everyone I interviewed had stories to tell. Some of the stories are specific to nuclear power, but, as you see in the following, their struggles were more generally about doing activism in Japan, a country that claims to be a democracy. As someone who has lived in the United States and Japan and is fluent in both English and Japanese, Smith offered interesting insights into Japan's democracy. She felt that the Japanese people are socialized to think democracy primarily in terms of voting. Additionally, sometimes labor unions or interest groups would visit their representatives to ask for certain legislative actions. But that would be the extent of participation in democracy. Smith observed that the Japanese read a lot about civil rights issues and are knowledgeable about them, but very few act upon the knowledge:

> There are books criticizing people in power, and you read about it. It's a democratic society, so these books exist and you are free to read them, and you read them. But it's passive if you don't take any action. When you take an action beyond voting, you are considered a highly unusual person. The other thing with democracy, if you ask people about MLK or Gandhi, everyone has a very high opinion. These two guys got arrested repeatedly. And yet sort of like they are praised, and yet if anybody was active and arrested, they would be treated like a pariah. It's sort of like a double standard.

This assessment of Japan as a passive democracy that discourages actions was shared by others I interviewed. Fujikawa felt that expressing one's view that is different from the crowd is not embraced: "I think there is a strong tendency to see political topics as taboos. For example, it is not easy to express your opinions in local community meetings. So, we all censor ourselves." Sato echoed Fujikawa. He felt that this taboo is perpetuated at school and home: "If you do bring it up, you would be told, 'why are you thinking about such a peculiar thing? You shouldn't worry about that!' So, there is this unsaid agreement that you should not talk about politics."

If you are vocal, you will stick out and can be seen as a radical or fanatic person. Sato noted that, before the Fukushima accident, few people paid attention to the activities of Greenpeace, and the organization was even widely misunderstood as a terrorist organization. Smith's comment confirms Sato's experience. Based on her many international experiences, she made the following observation:

> The gap is so huge between activism being honored abroad and activism not being honored here in Japan. The gap is unbelievable. After 3/11, maybe. Before that, certainly not. My daughter couldn't go out and hold her head up high and say "guess what? My mother is an activist." You know, the reaction would not be good. You can't say it in 99.9% of your social circle, you know. Even now, it's kind of like that.

Thus, the frustration resulted from the clash between the activists' view of public voice in a supposedly democratic country and the pervasive cultural norm that discourages political self-expressions and social activism and instead encourages self-monitoring, chastising, and silence over social and political issues.

The challenges also came from the legal constraints and law enforcement. Sato explained that, although the freedom of expression should be protected, people often get arrested simply by straying a little from the designated demonstration area and spilling out to the road. Having lived abroad and participated in demonstrations in other countries, Sato had the following comparison to share:

> In Germany, citizens have participated in social movements in massive numbers over the years and have protected freedom of expression. In America, there are occupy movements and you stay there until you are forcefully removed. You may be arrested, but you would be released before too long. In Japan, you would be jailed for a minimum of 20 days or so. Then, you would lose your job, and the impact of your activism on your family could be devastating. In the West, you can participate in a demonstration as though you are going to a parade. We don't have that here in Japan yet. That is a big barrier that social movements face here.

He felt that severe punishments for seemingly trivial violations deter the majority of people from expressing their thoughts more freely.

Smith brought up another kind of legal bias that specifically affects antinuclear activism. She noted that there are many lawyers who are committed to helping citizens bring lawsuits against nuclear power and offer their help practically pro bono. This makes it affordable for common citizens to take their cases to court. However, the legal system favors the government and utilities, making it extremely difficult for antinuclear lawsuits to see a victory even in cases where plaintiffs clearly have compelling arguments.

Marginalization of pre-Fukushima antinuclear activism

In the aftermath of the Fukushima accident, antinuclear protests were widespread. This was a drastic change from that which the veteran activists experienced before the accident. According to Yanagida, for two decades, members of antinuclear organizations in Tokyo came together to regularly hold rallies:

> There were about 20 of us from five groups. We would carry a banner saying "We don't need nuclear power. We have enough electricity without it!" and walked through the streets. We would carry 100–200 copies of flyers and tried to give them to people. From outsiders, it probably looked like a cult. Now [after the Fukushima accident] people would believe us, but, back then, even our friends didn't take us seriously.

Yanagida's experience of being belittled was echoed by others. Some also experienced more hostile reactions from the public. Murata, for example, shared repeated experiences he had while trying to collect signatures for a nuclear power matter:

> If people don't want to sign it, that is fine. But there are people who become verbally abusive or aggressive. For example: "Are you stupid?" "Why don't you open your eyes to the world?" "So, what are you going to do if we run out of electricity?" They are exposed to the commercial, pronuclear media day in and day out, so they begin to believe that it is their idea as well.

These experiences were psychologically draining to Murata. The examples from Yanagida and Murata give glimpses into the hegemony of nuclear power. It is so deeply embedded in the culture of Japan that questioning its existence was almost absurd. In the urban areas where nuclear power facilities are not an empirical reality, activists faced the challenge of having to make the threat real to the public, which was almost impossible when the utilities and the government had virtually unlimited resources to influence public discourse. As discussed in Chapter 4, they utilized a variety of media campaigns and public relations strategies to make nuclear power look safe, green, efficient, and even fun.

In *Fresh Currents*, a publication that addresses the road that led to the Fukushima accident and Japan's energy future, Smith (2012) wrote that her fight against the pronuclear lobbying was lonely. When I asked her about this in our interview, she responded that this loneliness is "almost chronic" because it was "like a taboo." She elaborated:

> Nuclear power is supposed to be o.k. and you are saying it's not. It would just make things uncomfortable. I would self-censor myself and wouldn't talk about it because it would make other people uncomfortable. . . . But everybody kind of knew, living in the same building for 28 years. They see me quoted over the years. But I have not had a single comment even to date from the people living in the same building.

Hence, perceived general cultural apathy toward social activists, the rigged legal system, and marginalization of antinuclear voices all made it hard for the activists to mobilize people around any social issues but particularly nuclear power. They felt that the media was partly responsible for creating the unhospitable public sphere for social activism in general and more specifically antinuclear movements.

Media as a barrier to antinuclear movements

I have already discussed the indispensable role mass media played in nurturing the positive framing of nuclear power, including newspapers (Chapter 3) and other forms such as television commercials and children's programs

(Chapter 4). Unsurprisingly, the veteran activists were critical of the media's coverage of nuclear power issues. One common area of complaint was the coverage of antinuclear power movements. Although the movements existed for decades, they were invisible because of the lack of media reporting. Murata pointed out the scarcity of news media reporting on nuclear power issues:

> Before Fukushima, I don't remember much media coverage of nuclear power. If there is no coverage, you are not informed, and you can't form your opinion. The media largely ignored the activities that bring negative attention to nuclear power. When something happened at a nuclear facility, you would go to the government to demand more information or ask their plan to address the problem. These are very important activities. But they don't get reported by the media.

Smith gave a few examples that illustrate Murata's observation, including the case in 1999 when her organization, Green Action, and Koyama's Mihama-no-kai became the plaintiffs to prevent the use of MOX fuel at the Takahama Nuclear Power Plant in Fukui Prefecture owned by KEPCO. Despite all the work their organizations did to force the company's accountability for its fraudulent data regarding MOX, KEPCO took the credit:

> Kansai Electric said that they realized that this problem existed. Falsification of the quality control material. They announced that they decided not to use MOX, and they are the ones in the lead headline of the front page the next day. And there wasn't a single line about how it happened. We forced them into that, but we were totally invisible. And the national headline news said, Kansai Electric found a problem, decided to correct it, not use the fuel, not to go with the program. This kind of thing happens over and over.

The activists felt that the failure of the conventional media goes beyond neglecting the work of antinuclear power movements but b; when they did report, the coverage tended to be biased against the movements. Sato observed that the skewed framing of antinuclear power movements reflects a deep-rooted pattern of media's treatment of social movements and social justice protests:

> If people get arrested for protesting something, the news media only say that they got arrested for violating this or that. Newspapers, for example, would simply put that violation news in the crime section, but they don't cover it as societal and political news. There is very little coverage of why they were protesting. Japanese news media rarely cover social movement stories in depth and listen to both sides of a conflict. So, the

audience doesn't get an accurate picture of what really happened. Demonstrators' stories are rarely covered.

In the activists' experience, therefore, the antinuclear power movements suffered both from the lack of coverage and biased portrayals.

Koyama observed that newspapers stopped covering social movements in general since the late 1980s, even when people surround the National Diet building in protest. He attributed the silence to the financial pressures that newspapers constantly face; for the media to survive, they must rely on private money of the industries that the movements criticize. In the case of nuclear power, he believed that newspapers' finances are closely tied to the nuclear power industry. This is of course not unique to Japan; commercial medias are owned and controlled by corporations, and they are not inclined to report on social movements that threaten corporate interests (Stewart et al., 2012). Financial ties may not be the only explanation, however. In their historical analysis of Japan's contemporary social movements, Chiavacci and Obinger (2018) observe that Japan's intense protest cycle ended around the mid-1970s for compounding economic, military, social, and political reasons. One of the reasons is delegitimization of confrontational movements. Until the 1970s, the media provided social movements an important public platform to challenge the conservative political leadership. However, as student movements became more violent and some extreme left groups committed violent actions, the media reporting became increasingly negative and stigmatizing to social movements in general (Chiavacci & Obinger, 2018). As media framing shapes public opinions (Entman, 1993; Entman & Rojecki, 1993), this also explains the widely held assumption of Greenpeace as a terrorist organization that Sato mentioned. Framing has consequences.

Traditional media's bias against social movements has been studied extensively as a phenomenon called the "protest paradigm." Media that operates within this paradigm serves a social control function by giving no coverage or negatively covering the groups or actions that threaten the status quo (Boyle, McCluskey, McLeod, & Stein, 2005; Boyle & Schmierbach, 2009) and by focusing more on the spectacle, action, and tactics rather than providing the substance of the social protest (Detenber, Gotlieb, McLeod, & Malinkina, 2007). From the point of view of the veteran antinuclear activists, the Japanese traditional media closely followed the protest paradigm, which led to further consequences; the audience remained not only ignorant but also developed negative views of the work the antinuclear power organizations and activists do.

Besides the skewed media coverage of antinuclear movements, the activists agreed that the media representations of nuclear power are biased in favor of the nuclear industrial complex. As shown in Chapter 4, utilities have used a variety of media outlets to advance a pronuclear narrative.

Having to fight this narrative regularly, Smith had a lot to say about this challenge. One articulation she perpetually combats is "nuclear energy is cheap," a widespread message accepted by citizens, including politicians and intellectuals:

> They used theoretical figures . . . assuming that nuclear plants oper-
> ated at 80% capacity all the time. Also they didn't include subsidies,
> that 80% of the public subsidies went to nuclear alone. . . . Also, when
> nuclear generation increased, they needed to do "pump storage" where
> you pump up water for later use when the demand is high. Because it's
> pumping up water, the nuclear industry made it look like it was part of
> hydroelectric system. That should be nuclear cost. When you subtract
> all that and make it look as if it's operating always at 80%, then, you
> come out really cheap, whereas the reality was that the taxpayers, the
> consumers are paying. . . . We are all taxpayers and consumers, so the
> public was paying for the subsidizing, but wasn't aware of it.

Another articulation Smith brought up was "nuclear is clean":

> They used to heavily advertise that nuclear produces no CO_2, which
> is not true. After the Kyoto protocol, this was heavily advertised. All
> the time. All the time. So, advertising that nuclear is green. Good for
> the climate. This has been used a lot. . . . But it won't help with global
> warming . . . [because] the uranium process creates a lot of CO_2 emis-
> sions. We worked on this issue, but we got no media coverage. Nuclear
> is actually bad for global warming.

Indeed, her organization has been addressing the flaws of the "green and clean" narrative at least since the mid-2000. Green Action has posted information on their website, organized seminars, and presented at multiple events. Its post from May 23, 2005, for example, is titled "Is nuclear power an effective solution for global warming?" and exposed the myth of the "zero emission" claim the government used in its Framework for Nuclear Energy Policy (Green Action, 2005). Despite the efforts of Green Action and other no-nuke organizations to communicate counter-narratives to the public, the extent of their reach was limited in the absence of the coverage by the conventional media. In the pre–social media world, the silence and skewed, favorable coverage of nuclear power by the mainstream commercial media, were particularly detrimental to the antinuclear power movements.

Conclusion

Urban antinuclear power movements roughly began in the 1970s when antinuclear power activists around the country gathered in Kyoto. Since then, the movements have been relatively invisible to the public until the

Fukushima accident, with the exception of the aftermath of Chernobyl. Through interviews with veteran activists, this chapter identified several compounding forces that contributed to the invisibility, including normative passive citizen participation in democracy, marginalization of antinuclear power movements, and media and legal biases. The activists' experiences of engaging in antinuclear power activism reveal insights into pre-Fukushima Japan's democracy and civic society.

Critical theorist of modernity Jürgen Habermas (1992, p. 367) described civic society as being "composed of those more or less spontaneously emergent associations, organizations, and movements that, attuned to how societal problems resonate in the private life spheres, distill and transmit such reactions in amplified form to the public sphere." The strength of civic society, however, varies locally. Aldrich (2008) argues that relatively autonomous groups in each locality "can support *or* contest state initiatives, making it less or more difficult for states to enact policies" (p. 15). In Chapter 5, we saw cases that demonstrated perseverance of local civic society that made it difficult for the private-state interests to enact the national nuclear power policy in specific local contexts. However, building a strong civic society in an urban context can be more challenging, especially when the problem it addresses is vast, technical, and already integrated into the fabric of the society. Nuclear power is particularly hard to address because it has the full backing of the powerful industrial complex, consisting of the utilities and the government, the conventional media and academics.

The media is undeniably essential to social movements as it provides visibility and shapes public opinion. Before the Internet became a major medium of communication and before social media became readily available, the conventional media played a far more powerful role in shaping the urban antinuclear power movements. Unfavorable media coverage of a social movement discourages citizens who are not part of the movement from forming an informed judgement, because they cannot formulate fair understanding of the changes the movement demands (Entman & Rojecki, 1993). *Lack* of media coverage has a similar consequence. Both silence and bias contributed to the challenges experienced by the veteran antinuclear activists, making it difficult for them to mobilize the civic society on state policies.

While the urban antinuclear movements faced many externally imposed challenges, it is important to acknowledge before closing this chapter the weaknesses of the movements that were brought up by one of the interviewees. Smith observed that both the local and urban movements focused on safety issues in opposing nuclear power, but not much effort went into economic analysis to refute the industry's claim that nuclear is inexpensive. While enormous subsidies financed by taxes supported the nuclear power industry, not enough attention was given to it as a way to oppose nuclear power. Similarly, Smith felt that not enough effort was made to fight against the claim that nuclear power is necessary for fighting global warming. Indeed, the history of antinuclear power movements in Japan paid almost exclusive attention to safety. Would the movements have been more visible if they included these other frames? We will never know the answer. However, we can reasonably imagine that the use of a variety of frames and making connections between the frames could have potentially expanded the civic society. Coalitions play a critical role in social movements, but groups will not work together unless they share some common goals (Van Dyke & McCammon, 2010). Strategic use of frames can encourage coalitions and alliances between groups that may not work together otherwise. Despite the weaknesses and relative invisibility, however, the importance of the urban movements should not be understated. Their perseverance and decades of work laid the foundation for post-Fukushima antinuclear power movements we discuss next.

Notes

1 This is not unique to nuclear power issues, but Japan's civic society in the form of social movements existed rather invisibly from the mainstream society. These decades were a period of abeyance in which thousands of small social movements, a mix of old and new group addressing a variety of social and political issues (Steinhoff, 2018).
2 Eri Fujikawa is a pseudonym.
3 Hiro Murata is a pseudonym.
4 Many of these studies are traced back to Geertz Hofstede's theory of cultural dimensions resulted from his study of comparing national cultures in organizational contexts in the early 1980s.

References

Aldrich, D. P. (2008). *Site fights: Divisive facilities and civic society in Japan and the West*. Ithaca, NY: Cornell University Press.

Ando, T. (2019). *Datsugenpatsu no undoshi [Antinuclear power movement history]*. Tokyo: Iwanani Shoten.

Angelique, H. L., & Culley, M. R. (2014). To Fukushima with love: Lessons on long-term antinuclear citizen participation from Three Mile Island. *Journal of Community Psychology*, 42(2), 209–227. https://doi.org/10.1002/jcop.21605

Boyle, M. P., McCluskey, M. R., McLeod, D. M., & Stein, S. E. (2005). Newspapers and social protest: An examination of newspaper coverage of social protest from 1960 to 1999. *Journalism & Mass Communication Quarterly*, 82, 638–653. https://doi.org/10.1177/107769900508200310

Boyle, M. P., & Schmierbach, M. (2009). Media use and protest: The role of mainstream and alternative media use in predicting traditional and protest participation. *Communication Quarterly*, 57(1), 1–17. https://doi.org/10.1080/01463370802662424

Chiavacci, D., & Obinger, J. (2018). Towards a new protest cycle in contemporary Japan? The resurgence of social movements and confrontational political activism in historical perspective. In D. Chiavacci & J. Obinger (Eds.), *Social movements and political activism in contemporary Japan: Re-emerging from invisibility* (pp. 1–23). New York: Routledge.

The Citizen's Nuclear Information Center. (n.d.). Retrieved from https://cnic.jp/english/?page_id=13

Detenber, B. H., Gotlieb, M. R., McLeod, D. M., & Malinkina, O. (2007). Frame intensity effects of television news stories about a high-visibility protest issue. *Communication & Society, 10*(4), 439–460.

Entman, R. M. (1993). Framing: Toward clarification of a fractured paradigm. *Journal of Communication, 43*(4), 51–58. https://doi.org/10.1111/j.1460-2466.1993.tb01304.x

Entman, R. M., & Rojecki, A. (1993). Freezing out the public: Elite and media framing of the U.S. Anti-nuclear movement. *Political Communication, 10*, 155–173.

Green Action. (2005, May 23). *"genpatsu wa ondanka ni yuukou" nanode shouka [Is nuclear power an effective solution to global warming?]*. Retrieved March 9, 2021, from http://greenaction-japan.org/jp/2005/05/221

Greenpeace Japan. (2012). *Take action for positive change!* Tokyo: Greenpeace Japan.

Habermas, J. (1992). *Between facts and norms.* Cambridge, MA: MIT Press.

Honda, H. (2019). [Review of the book *Datsugenpatsu no undoshi*, by T. Ando]. *The Ohara Institute for Social Research, 734*, 94–97. http://oisr-org.ws.hosei.ac.jp/images/oz/contents/734_07.pdf

Jesty, J. (2012). Tokyo 1960: Days of rage & grief. *MIT Visualizing Culture.* Retrieved from https://visualizingcultures.mit.edu/tokyo_1960/anp2_essay06.html#n10

Nishio, B. (2013). *Watashino hangenpatsu kirinukicho [My antinuclear power scrap book].* Tokyo: Ryokufu: Shuppan.

Packard, G. R. III. (1966/2015). *Protest in Tokyo: The security treaty crisis of 1960.* Princeton, NJ: Princeton University Press.

Smith, A. M. (2012). Post-Fukushima realities. In E. Johnson (Ed.), *Fresh currents* (pp. 79–88). Kyoto, Japan: Kyoto Journal/Heian-kyo Media. Retrieved from http://download.freshcurrents.org/

Steinhoff, P. G. (2018). The uneven path of social movements in Japan. In D. Chiavacci & J. Obinger (Eds.), *Social movements and political activism in contemporary Japan: Re-emerging from invisibility* (pp. 27–50). New York: Routledge.

Stewart, C. J., Smith, C. A., & Denton, R. E. Jr. (2012). *Persuasion and social movement* (6th ed.). Long Grove, IL: Waveland Press.

Sueda, K. (2018). *Kusanone.* Retrieved from http://ksueda.eco.coocan.jp/kusanone1807.html

Van Dyke, N., & McCammon, H. J. (2010). Introduction: Social movement coalition formation. In N. Van Dyke & H. J. McCammon (Eds.), *Strategic alliances: Coalition building and social movements* (pp. xi–xxviii). Minneapolis: University of Minnesota Press.

Yamamoto, M. (2004). *Grassroots pacifism in post-war Japan: The rebirth of a nation.* New York: Taylor & Francis.

7 Fukushima and (re)claiming the voices of democracy

The Fukushima Daiichi nuclear disaster in March 2011 was the most serious nuclear power accident in Japan and was one the two largest in the world's nuclear power history. The accident and the state's poor handling of the aftermath became an antagonism to nuclear hegemony, mobilizing the largest social movement Japan saw in almost a half century. The veteran activists, who experienced a number of challenges in engaging in antinuclear activism, suddenly found themselves in the company of tens of thousands of citizens who participated in public demonstrations for the first time in their lives. The traditional news media, a historically pronuclear apparatus, could no longer ignore the widespread protests. Yet their coverage continued to be thin from activists' point of view. On the other hand, the spread of the Internet and social media allowed citizen journalism to document the gaps. This chapter explores various ways in which Fukushima became a vehicle for citizens to discover and exercise their democratic voice. I develop this chapter based on published sources (books, scholarly papers, newspaper articles, and websites) and my own fieldwork of participating in the sites of antinuclear power movements. Out of many sites of antinuclear power movements, I pay particular attention to three sites: the rallies in metropolises, including periodic large demonstrations and the weekly protest rallies in Tokyo; the occupy-style movement that occurred in Tokyo; and the protest against the restart of a reactor after a period of zero-nuclear power that lasted two months. These sites illustrate varied expressions of post-Fukushima antinuclear power movements and more generally the changing face of social movements and their role in civic society.

Rallies in metropolises

One of the first rallies was organized in Tokyo a month after the disaster by *Shirōto no Ran* (amateur's riot), a loose network of individuals who call for alternative economy and more public participation in democracy. They used existing networks and social media to advertise the rally and drew over 15,000 people to a demonstration full of music and dance (Steinhoff, 2018). On the three-month anniversary of the disaster, thousands of protesters,

DOI: 10.4324/9781003044222-7

including students, teachers, parents with children, and workers, gathered to express their anger over the government's inadequate response to the crisis by chanting "No Nukes!" and "No more Fukushima!" (Slodkowski, 2011). Then, in July, about 1,700 people, including those who evacuated from the area around the damaged nuclear power plant, attended the rally in Fukushima City organized by the Japan Congress Against Atomic and Hydrogen Bombs (France 24, 2011). One of the largest rallies, attended by 60,000 people, was held in Meiji Park on September 19, 2011. Since then, several more relatively large-scale demonstrations occurred in Tokyo in the next year as part of the *Sayonara genpatsu 1000 mannin akushon* (goodbye nuclear power 10 million citizen action). The movement aimed to collect 10 million hand-written signatures to call an end to Japan's nuclear power. This movement was launched by nine public intellectuals, including the Nobel Laureate Kenzaburo Ōe, composer Ryuichi Sakamoto, and Buddhist nun Jyakucho Setouchi. Aside from the interruptions by the pandemic, the movement has held periodic rallies to renew their commitment to the cause, although the number of participants has fallen over the years. On the ten-year anniversary of Fukushima in March 2021, they organized a rally at Hibiya Open Air Concert Hall in Tokyo.[1] A singer opened the rally followed by a series of speakers, including Naoto Kan, former Prime Minister (PM) who came out to oppose nuclear power after Fukushima. Following the speeches, about 1,500 people in attendance marched the streets of Tokyo. The route included the headquarters of TEPCO, the owner of Fukushima Daiichi Nuclear Power Station. The speakers' messages were unequivocal; Japan must abandon nuclear power, and Fukushima is not over as there are still over 40,000 people who cannot return to their hometowns in Fukushima (Sayonara-nuke, n.d.). They have collected over 8.8 million signatures as of March 3, 2021.

Another important piece in the post-Fukushima urban movement is the weekly Friday rally in front of the Prime Minister's residence (the PMR Friday Rally hereafter). *Shutoken Hangenpatsu Rengo* (Metropolitan Coalition Against Nukes or MCAR), a network of antinuclear organizations that formed in September 2011, began to hold rallies, starting in April 2012 (Brown, 2018). Initially starting small with just a few hundred participants, the PMR Friday Rally swelled in number after the announcement of the resumption of the nuclear power program in June 2012. In the mid-June, then PM Yoshihiko Noda ordered the restart of No. 3 and No. 4 reactors at the Ohi Nuclear Power Plant in Fukui Prefecture and set July 1 as the resumption date for the first reactor. The citizens responded immediately. In Tokyo, MCAR acted swiftly to organize demonstrations outside the Prime Minister's residence. According to Yasumichi Noma (2012), one of the organizers, the first demonstration on June 20 was attended by 40,000 people, and the following week the number swelled to 200,000.[2] The citizens were already livid about the repeated poor handling of the Fukushima disaster and its victims by the government. Polls showed that two-thirds of the citizens felt that nuclear power plants are dangerous (Fackler, 2012). Noda's announcement poured more fuel to the fire of outrage. Similar weekly Friday demonstrations occurred in Osaka City, Kyoto City, and other larger cities around the country.

In 2013 when the demonstrations were robust, I was able to attend the PMR Friday Rally during my trip to Japan. Nearing 6 p.m., people began to appear from every direction and filled the sidewalks around the Prime Minister's residence. They represented diverse demographics: elderly women and men, young mothers with children, middle-aged men in business suits carrying briefcases. They held homemade signs with a variety of messages such as "no nuke," "we don't need nuclear power," "protect children, protect the Earth, end nuclear power now," "Tokyo Electric Power Company and the government are guilty," "Give us back Fukushima," "Decommission all reactors," and "Don't export nuclear power." There were also several people who participated in the rally on their bicycles with banners on their back. The organizers had drums and led chants. Besides the sheer number of participants (the lines seemed endless), what caught my attention was the artistic creative displays of antinuclear sentiments. A woman in her 60s, a regular at the PMR Friday Rally, for example, carried a yellow

Figure 7.1 The PMR Friday Rally.

Source: Photograph by the author.

protest umbrella with no-nuke messages printed on it. From its frame hang signs indicating the hazards of radioactive elements of nuclear fission such as cesium 137 and strontium 90. Along the sidewalk where protesters gathered, there were cartoon drawings, mocking the political leaders and warning the audience not to trust the politicians. Activists with artistic talents would bring these signs for the weekly rally for everyone to see – the demonstrators, bystanders, and the police. Another artifact caught my attention. An elderly man held in his hand the Japanese flag, which read, "clean nation." This intrigued me, because the Japanese hardly ever display the national flag unless they are cheering athletes in international games. Otherwise, the flag is often displayed by ultra-conservative, nationalist groups. The man told me that he wanted to carry the flag because a protest for a better Japan is a patriotic thing, and he wants Japan to transition to renewable energy. With a group of people I met earlier, I stood for the first time in front of the PM's residence and listened to the words of the organizers that were blasted through a megaphone. All this liveliness was juxtaposed with the expressionless faces of the police officers that were there to control the crowd. There were also streams of people who simply passed by the demonstrators, perhaps on their way home at the end of the day, trying not to make eye contact with any demonstrators. The chants and individual voices of objection formed cacophonous rackets loud enough for the people inside the walls of the government buildings around to hear. I wondered if our representatives in the buildings are affected at all by the weekly rallies that drew thousands of people.

Despite the mass rallies in cities and the extraordinarily intense protests at the Ōi site (a subject I discuss later), the reactors were restarted in July, ending the 15-month nuclear-free period since the meltdown. The movement did not take this as a defeat but demanded a meeting with Noda. Noda initially refused on the grounds that there is no precedence for such a meeting, but he eventually agreed to meet with the group upon urging by former PM Kan.[3] The leadership of MCAR met with Noda in August 2012 and demanded the government to stop the Ōi reactors and to decommission all reactors ("Sōri to hangenpatsudantai," 2012, August 22). Noda responded that the restart of Ōi was a decision that was made after obtaining the assurance of the reactors' safety and that the government's basic policy is to end reliance on nuclear power. The meeting lasted for 30 minutes and made no difference to Japan's nuclear power policy. The value of this meeting, however, may rest elsewhere. It is noteworthy that this was the first time that the PM met with an antinuclear power activist group. In Japan's representative democracy, citizens have virtually no opportunity to directly address their PM. This meeting broke that norm; "no precedence" is no longer an excuse the head of the nation can use.

Here, we must acknowledge the role of Naoto Kan. Kan was the PM (2010–2011) at the time of the Fukushima nuclear accident. This accident had such a profound impact on him that, while he was in the office, he

Figure 7.2 Drummers setting the beat for the chants at the PMR Friday Rally.

Source: Photograph by the author.

Figure 7.3 A protester on a bike at the PMR Friday Rally.

Source: Photograph by the author.

Figure 7.4 Antinuclear poster displayed at the PMR Friday Rally.
Source: Photograph by the author.

declared that Japan would phase out the nuclear power program. In his interview with *Deutsche Welle* (DW), he stated, "The accident radically changed my perspective. I now consider nuclear energy to be the most dangerous form of energy, and the risks associated with it are too great for us to continue generating atomic power" (Dominguez, 2015). After his resignation in the fall of 2011, he became even more vocal about ending the program and the need to rapidly transition to renewable energy. His appearance at the 10-year Fukushima anniversary rally and his role in arranging the meeting between MCAR and Noda represent the activist side of this politician. Kan is not alone; Fukushima flipped other pronuclear politicians, including two former PMs, Junichiro Koizumi (2001–2005) and Yukio Hatoyama (2009–2010). These three PMs attended a ten-year anniversary no-nuke event organized by a community organization (Okuhara, 2021). Two other former PMs, Morihiro Hosokawa (1993–1994) and Tomiichi Murayama (1994–1996), also sent messages supporting the end to nuclear power. During his speech, Hatoyama asked Koizumi to return to politics so the three of them can establish a no-nuke political party. It is not clear if this was said to simply demonstrate the spirit of solidarity or whether the idea of a new political party is under serious consideration. Regardless, the antinuclear voices of the powerful politicians, unprecedented before Fukushima, illustrate that proponents no longer have the monopoly in the political discourse of nuclear power.

The PMR Friday Rally lasted for ten years and ended with its 400th event on March 26, 2021. After the initial participation of tens of thousands of people, the movement continued to mobilize thousands until about 2015, but, as the number of participants declined over time to mere some dozens, it became difficult to sustain the movement that relied on the donations from participants (Saito, 2021). The organizers stated that the PMR Friday Rally would stop for now, but MCAR would continue to exist until the government decides to end the nuclear power program (Usui, 2021).

Photojournalist Takatomo Usui (2021), who attended the protests since 2015 to document the movement, observed that the PMR Friday Rally represented "a new grass-roots, democracy movement"; the protest represented "a new tool of democracy – a direct political action." However, as discussed in the preceding chapters, large-scale direct political actions occurred in the 1960s and 1970s (e.g., *Anpo* protests, student movements, labor movements). What was then different about the PMR Friday Rally and other post-Fukushima rallies? Recall Smith's comment in Chapter 6 that Japanese citizens are educated but lack actions. The large-scale rallies afforded a comfortable space for ordinary citizens to participate in collective self-expressions about a national policy. One of the rally participants, a woman in her 70s, that Usui interviewed stated, "I had never spoken publicly about politics or even expressed my opinions. The Friday rally taught me that it is o.k. to express your opinion." Similarly, a 60-year-old man who regularly attended the Friday protest since 2013 said in another interview,

> I felt it was important to act after that horrific accident. Until then, I never participated in any demonstration, because I felt that it was not right to challenge the government. But it was easy to participate [in the Friday protest] and raise my voice.
>
> (Saito, 2021)

Direct actions were therefore not new to the post-Fukushima demonstrations because these actions had already existed. Rather, as the testimonies suggest, it is the ease of participation that makes them different from the prior protests. Partaking in protests became ordinary and psychologically accessible. In Chapter 6, Sato from Greenpeace Japan observed the ease with which Americans and Germans seem to attend demonstrations – something that, he felt, was lacking in Japan. The post-Fukushima movements brought this psychological accessibility to Japan.

No nuke tento Hiroba

In addition to these classic displays of activism through demonstrations and rallies, the post-Fukushima urban antinuclear movement also took an occupy form. On September 11, 2011, the six-month anniversary of the accident, a *datsugenpatsu tento* (no-nuke tent) went up right in front of the METI. It was built by antinuclear, peace activists from Tokyo following a human chain protest that surrounded the METI building. As METI has jurisdiction over the country's energy policy, erecting a tent in this space was an unequivocal challenge to the policy that antinuclear activists saw as responsible for the accident. The activists argued that the space should belong to the citizens for the debate over nuclear power (Brown, 2018). A month later, another tent went up right next to the first one. This time, it was built by a group of women from Fukushima. The Women's Tent, as it came to be called, was a result of over 100 women from Fukushima staging a sit-in in front of the first tent. The women were emboldened by the audacious actions taken by the activists to build the tent and decided to join them to protect the space and call for justice for people in Fukushima (AM Kikaku, 2012). Together, two tents became what investigative writer Satoshi Kamata called "Kasumigaseki no heso" (the belly button of Kasumigaseki) (AM Kikaku, 2012). Kasumigaseki is a district where most of Japan's cabinet offices are located. It is not an area of Tokyo ordinary citizens visit. After the tents occupied a corner space in front of the METI building, Kasumigaseki no longer belonged to bureaucrats alone. The corner space became a people's plaza where anyone could and would visit. The no-nuke tents claimed it as a space where citizens could gather and participate in democratic expressions. The space came to be called *Keisanshōmae Tento Hiroba* (no nuke tent plaza in front of the METI) (*Tento Hiroba* henceforth).

Interestingly, in the eyes of the Japan Postal Service, the tent plaza became an official address even though the METI dismissed the legitimacy of the plaza. The occupants could receive mails and packages. They were able to set up a bank account using the address to receive donations. The tent activists told me that they received many new year cards at the beginning of that year. Symbolically, this is interesting because new year cards are an important Japanese cultural tradition where people exchange greetings, wishing their friends, family, and colleagues a healthy, prosperous, and happy year. People at the tent plaza received cards from strangers all over Japan who expressed gratitude and support for their activism. As the happenings at the *Tento Hiroba* are shared through its website, social media, and other online sources, material support also poured in, including water, food, blankets, donations, and more.

The organizers of the *Tento Hiroba* initially thought that the authorities would forcefully remove the tents in a few days. The METI posted signs around the tents, stating, "National land. Do not enter. Permitted personnel only." Its security guards would also ask the occupants to leave. However, they did not physically remove the occupants or the tents. After 200 days of occupation, one of the organizers, Tadao Eda, speculated that the METI had not been able to remove them because the *Tento Hiroba* represented the citizens' sincere resolve (AM Kikaku, 2012). Moreover, the police could not remove the tents or arrest the occupants because they did not have

Figure 7.5 Protesters and visitors gathering in front of the two tents at the Tento Hiroba.

Source: Photograph by the author.

Figure 7.6 Inside of the Women's Tent at the Tento Hiroba.

Source: Photograph by the author.

jurisdiction over the national land where the tents stood. It was March 2013 when the METI (representing the Japanese government) finally brought a lawsuit against the two representatives of the no-nuke tent on the ground of illegal occupation of the national land. In February 2015, the Tokyo District Court ruled in favor or the METI. In July 2016, the Supreme Court rejected the activists' appeal and ordered them to remove the tents. As the activists refused to do so, the tents were forcefully removed on August 21, 2016. The two large no-nuke white tents survived for five years and became a conspicuous symbol of the post-Fukushima antinuclear movement. After the removal of the tents (and to this day), the activists have continued their activities, including regularly demonstrating at the same corner and publishing a daily journal.[4]

What role did the *Tento Hiroba* play in the overall movement? One of the Fukushima women who set up the women's tent, Chieko Shiina, called the tents *norishiro* (the margins of papers where you apply glue to piece together two papers) because it is a space where people meet and connect. As a Fukushima resident, Shiina particularly saw the tent plaza to keep Fukushima alive in the minds of people:

> The tents bring people together. Of course activists come, but others come, asking what they can do. A strong network has emerged through the tent. It is a space where we come together to renew our pledge that we won't let the government forget about Fukushima. It is a space where we bring news from Fukushima and share with others. It is a space where we think about Fukushima.
>
> (AM Kikaku, 2012, pp. 63–64)

In fact, the Women's Tent offered a space to talk about Fukushima in more intimate, interpersonal ways rather than as part of a protest speech. For example, in my interview, Shiina shared that many people in Fukushima live with the fear of radiation poisoning but cannot even talk to doctors because they dismiss the possibility. She gave an example of her friend who went to see a doctor because she got ill and was afraid that radiation from the melted plant had something to do with it. The nurse there warned her that she should not mention "radiation" because the doctor may not see her if she does. As a result, the friend continues to live with an unresolved fear. The tent offered a space to share personal and emotional stories.

Although the Fukushima nuclear accident prompted the *Tento Hiroba*, it became a space for more than justice for Fukushima or even an antinuclear power movement. One of the regular occupants of the No. 1 tent, Tamotsu Shimoyama, saw the tents as a safe space where strangers can drop by and strike a conversation. For example, he shared an anecdote of sitting in front of the tents next to a 74-year-old woman who told him that she enjoys visiting the *Tento Hiroba* because she makes new friends, and the tents make her feel that it may be really possible to end nuclear power (AM Kikaku, 2012).

My experience with the *Tento Hiroba* confirms these testimonies. In June 2012 and again in June 2013, I had multiple opportunities to visit the *Tento Hiroba*. On my first visit, I went to one of the tents that had a little opening with some writing. The sign said "uketsuke" (reception). A slender woman with a short, bobbed hair greeted me and asked me to sign the "hou-monsha" (visitor) book as all visitors are asked to do so. She told me that over 10,000 people had already visited the tents. I explained the purpose of my visit, and she immediately took me to the next tent, the "Women's Tent" or "No. 2 Tent," telling me that someone is visiting from Fukushima and I would probably benefit from talking to her. Unfortunately, the Fukushima woman was gone to a meeting. I saw two women, one in her 30s and one in her 60s, in the tent, who immediately welcomed me. The women were eager to explain to me how the *Tento Hiroba* is run. The tents were run by volunteers who took turns being there. The first tent was occupied by someone 24 hours a day to make sure that the tents were not removed or vandalized. Volunteer men took turns sleeping there. The security of METI and police came regularly to harass and sometimes beg the volunteers in the tents to leave, but they had not done anything forcefully yet.

One of the women at the tent, Ms. Ōta, was particularly passionate about her participation in the movement and was one of the regular volunteers at the women's tent. She did not become active until after Fukushima, but her awakening to the nuclear power issues occurred a couple of years before the accident through the antinuclear information she came across at an event she attended. Until then, she never knew that there were as many as 54 nuclear reactors in Japan. When I asked her why she became part of the movement, she replied, "It is a matter of life or death. It's urgent. I feel that this is the last chance for Japan. If we don't wake up and end it, that will be the end of Japan."

The young woman next to Ōta nodded, "It's the same for me." While I was listening to Ōta, a woman in her 30s who evacuated to Kanagawa Prefecture from Fukushima over a year ago visited the tent. She told me that her 10-year-old son spoke at *Sekai taikai* (an international antinuclear conference) that happened recently in Yokohama. She herself went to the United Nation's headquarter in New York to explain the danger of nuclear power plants and to share her experience of living as an evacuee. If local activists who fought to stop the construction of a nuclear power plant were thrown into activism to protect their homes, these evacuees were thrown into activism because their homes were taken away. They would visit the tents because they could not stay inactive. But I wondered if the tents offered a bit of sense of home.

Soon, another visitor came in. A young British man, Thomas. He is a member of the Campaign for Nuclear Disarmament (CND) in England. He was living in Japan, teaching English, and just renewed his visa so he can stay for another three years. He stopped by the tent plaza to offer his help with the movement. It was not unusual for foreign residents and visitors to stop by the *Tento Hiroba* and offer help or words of encouragement, Ōta explained. Thanks to social media, the belly button of Kasumigaseki appeared to have extended its cord beyond the national or nationality boundary. People pop in and out, exchanging greetings, sharing what's happening in each other's life, chatting, and offering snacks. For a moment, it reminded me of how people are in my rural hometown; neighbors would drop by unannounced, have a cup of tea, and talk about what is going on in their lives.

When I returned to the *Tento Hiroba* the following summer, 2013, I received the same welcome from the volunteers. On the tatami mat, visitors and volunteers would sit and chat as friends would do. The conversations at the tent went beyond nuclear power, however. While I was there, the women covered a wide range of topics that they felt important: relearning how to be part of nature; listening to our gut and heart, not the television news; learning history from the point of view of people who did not have power; and fighting to protect Article 9 of Japan's Constitution. Today, there are very few opportunities in Japan where you meet strangers and engage in impromptu, sustained discussions about societal issues. The *Tento Hiroba* provided an extraordinary space of civic society to exercise the lost art of civic life – to have conversations with strangers.

Occupy Ōi

My second visit to the *Tento Hiroba* in 2012 occurred a few days after Noda's announcement of reactivation of the Ōi Nuclear Power Plant. I again went to the Women's Tent and found Ms. Shiina, one of the Fukushima women who launched the Women's Tent. Soon three others arrived. The conversations among the tent activists naturally concentrated on what can be done to stop Ōi's restart. They were frustrated and angered by the impending restart of the reactors when people in Fukushima lost their homes and still lived in fear. Two months ago in May when all reactors went offline, the antinuclear power activists around the county celebrated the occasion and hoped to keep Japan nuclear-free. Stopping Ōi's restart was critical. Shiina shared that networks of activists agree that something dramatic has to happen: staging a large die-in with all celebrity leaders like Kenzaburo Ōe, in front of the Diet building, for example. The men in the No. 1 tent are also pondering the idea to block the road to the Ōi facility with hundreds of cars. In any case, a protest at the facility was going to happen.

The day before the proposed protest, the *Tento Hiroba* chartered a bus so the activists from Tokyo can join the Ōi occupy protest that lasted 36 hours. According to one of the initiators of the protest, Shiro Yoshioka (2012), it is important to recognize the role of the protest tents that began to show up near the Ōi plant in April. The tent village, as they called it, grew to 25 tents a week before and then to 55 tents the day before the protest. They served as the local home for the Ōi movement. The protest began in the afternoon of June 30 with six cars barricading the road that leads to the plant to create a "freedom zone" between the cars and the gate of the main entrance of the nuclear power plant. Several activists tied themselves to the gate with chains and ropes. Protesters began to arrive, and hundreds of cars came. Nearly 700 people participated in the protest, chanting, drumming, singing, and dancing even though it was raining. It was women, men, old, and young – many mothers with small children. Although there were initiators, the protest was leaderless just as the rallies in metropolises were. Everyone was a leader, and no one was *the* leader. The leaderlessness confused *Jiētai*, the Self-Defence Force, that was deployed to control the protesters. To terminate the protest, the *Jiētai* police needed to identify the responsible party, but they could not because everyone was the responsible party (Sakata, 2012). Masako Sakata (2012), an activist who chained herself to the gate, recounted her experience:

> For 30 hours, we were non-stop. Everyone chanted, "saikado hantai!" (oppose the restart). We were free, cheerful but serious. Some were singing. Some were dancing. There were drummers. Some were giving flowers to the *Jiētai* officers. And some offered the officers umbrellas. Some chants were funny. But we did not withdraw from the barricaded area. It was different from the old kinds of protest that were scary. This was a new way to protest.

This did not mean that there was no physical conflict. When *Jiētai* began to push against the protesters, they pushed back. Soon, however, some protesters started to raise both of their hands to show they mean no violence, and others followed (Sakata, 2012). In a footage captured by a citizen photographer, Maiko Tsujimura (2012a), the protesters not only raised their hands but chanted "oppose violence!" It is not clear if it was directed to the police pushing against the protesters or if it was a declaration that their protest was peaceful. It may have been both. As I watched the video footages of the protest, I was struck by the lack of violence. The direct corporal protest sharply contrasted similar ones I have repeatedly seen on the U.S. news where peaceful protesters were often pepper-sprayed, battered, or in some cases shot.

Another participant, Uiko Hasegawa (2012), wrote that she was afraid that the Ōi protest was seen as too radical by the public and even by other antinuclear power activists, but it ended up drawing many regular citizens. She wrote about her experience of taking a part in a sit-in. In the late afternoon of July 1, the prefectural police announced through a megaphone that they would start removing the protesters. Following that, people began to sit and created human chains. Hasegawa was sitting toward the front with other women. They were all new to the antinuclear movement like her, let alone to a direct confrontation with the police. Everyone sat and linked their arms to avoid being removed by the police. About an hour later, the police began to physically remove the protesters:

> Five or six *Jiētai* worked on each protestor to forcefully unlink the arms, while other *Jiētai* prevented other protesters from trying to stop the forceful removal. The men protesters in front of us were all removed this way, and they came to remove our line. My heart was pounding with fear. The woman on my left was crying. But no one gave up. Then, it was just me and another woman in our line, and we tried to resist by hugging each other. But we both are petite, so they just carried us together.

Within a couple of hours, the police cleared the first several lines of sit-in activists and moved deeper into the "freedom zone" to remove more protesters. Those who were removed earlier returned to block the police. In the meantime, the protesters learned that the president of the KEPCO that owns the Ōi plant, the vice minister of the METI, and plant technicians used a boat to enter the plant from the seaside to officiate the reactivation of the No. 3 reactor. At around 2 a.m. of July 2, the participants collectively decided that it was important to end the protest without arrests or serious injuries. They concluded the 36 hours of resistance, clearing the barricades, cleaning the roads, and thanking the police (Hasegawa, 2012).

What have the movements accomplished?

What have these rallies, occupy tents, and the resistance at the nuclear power plant accomplished? In response to YouTube videos and web reports about the movements, I often came across variations of this comment: "these demonstrations don't change anything, neither the society nor the politics." It is true that the first reactor was reactivated as scheduled despite the protest at the Ōi plant and the demonstrations around the country. Other reactors followed. Under the new stricter safety standards, nine reactors at five nuclear power plants operate as of June 2021. Several other reactors are under inspection with the anticipation of being approved. It may be concluded that the post-Fukushima antinuclear movements made little difference from the policy outcome perspective at least for now. But policy changes are often slow. Japan as a culture needed to strengthen its protest culture as part of its civic society, and that is what these movements began to achieve.

Protests as embodied dis/articulations

If we look beyond the policy outcome, the post-Fukushima antinuclear movements made noteworthy differences. In his study of the waves of global protests that ensued following the 2008 financial crisis, Paolo Gerbaudo (2017) used the word *citizenism* to describe the nature of the protests. Citizenism is "the ideology of the 'indignant citizen' a citizen outraged at being deprived of citizenship, chiefly understood as the possibility of individuals to be active members of their political community with an equal say on all important decisions" (Gerbaudo, 2017, p. 7). A variation of this citizenism is at work in Japan's antinuclear movements. The movements are literally an embodiment of disapproval of the government and the utilities, of solidarity with Fukushima residents, of the values they want to uphold, and of desires for an alternative future. The demonstrations were the medium through which the citizens brought their thoughts, identity, and values into discursive presence. These representations matter because they make the material world meaningful and create possibilities (Hall, 1997). If the pre-Fukushima pronuclear discourse succeeded in naturalizing and legitimizing nuclear power as an indispensable, righteous national policy whose decision-making belongs to the authorities, then, the protest movements disarticulate this signification and rearticulate nuclear power and Japan's future. By so doing, the movement participants have articulated a different embodiment of citizenship. Recall the idea of passive democracy and the cultural tendency to suppress one's political opinions as observed by the veteran activists in Chapter 6. Ms. Shiina from the Women's Tent also brought up the cultural reason when explaining why she and other Fukushima women came to Tokyo to protest:

> Our culture discourages self-expressions. So, there is a very weak awareness about speaking as a right. Self-identity is weak. There is a lack of owning words. Because democracy is not something we had to fight for, people don't speak up even when human rights are violated. So, we had to rise up.

The post-Fukushima antinuclear movements, then, brought to citizens this awareness that speaking one's mind is a right. Ordinary citizens discovered that activism does not belong to select groups, but they can be inspired to act (Angelique & Culley, 2014). Hasegawa (2012) and the Friday rally participants alike discovered that raising their voices against government conduct they see unjust is their right. As the large numbers of ordinary people rose up, the movements changed the radical image of protests from the 1960s and 1970s and made them ordinary. Demonstrations have become part of the political culture of Japan since 2012; whatever the issue is, it has become commonplace to protest in front of the Prime Minister's residence or the National Diet building (Oguma, 2015). While protests can both threaten and advance democracy generally speaking (Arce & Rice, 2019), it is fair to say that the post-Fukushima movements have fostered greater active democracy by creating spaces for civic society to directly *and* comfortably express thoughts that were not fairly or fully accounted for in the representative democracy.

Protest media

The post-Fukushima antinuclear movements also impelled the traditional media to pay attention to social movements. During my visit in 2012, the tent activists were frustrated by the lack of media coverage of the antinuclear power movements. They complained that the Japanese news media rarely visited the tents whereas foreign media came frequently to broadcast the antinuclear power movement unfolding at the *Tento Hiroba*. One of the volunteers was a college student who had been helping a new organization that focuses on the well-being of Fukushima children. He was studying in England but was back to Japan for the summer. He felt that Japan's news media completely neglects demonstrations, which "makes it hard to keep the movement going. In contrast, if there is a demonstration with 100 participants in the United States, it will be on the news. Anti-establishment movements don't get media coverage in Japan, so citizens remain uninformed." He pointed out that what these conventional media outlets publish is important because most people in Japan still primarily rely on newspapers and television rather than the Internet for important news. Shiina agreed and commented that "Japan is still imperial when it comes to the media. People normatively believe that NHK is accurate whereas the information from the Internet is not trustworthy."

While the tent activists viewed the news media's lack of coverage as an intentional decision, sociologist Eiji Oguma (2015) offered a different insight. He closely followed the PMR Friday Rally and concluded that the lack of news media coverage of the demonstrations was due to structural flaws and the lack of sensibility to comprehend the value of demonstrations rather than a political reason. By structural flaws, he meant two problems. One is the gap in the responsibilities of journalists. There are journalists who are assigned to cover the politicians and bureaucrats, and they are stationed at the Prime Minister's residence and the National Diet building in Kasumigaseki. They see the massive demonstrations unfolding outside but do not think it is their job to cover them. And there are journalists who cover social issues, but they are not sent to Kasumigaseki. The second problem is the lack of experience in covering large demonstrations that are not organized by established organizations such as labor unions and political parties. Reporters do not know whom to interview and what viewpoints to collect and end up not covering the story. The lack of sensibility Oguma referenced has to do with both reporters and editors. Due to the lack of experience with mass protests, reporters may ask inapt questions. Even when reporters cover demonstrations and ask right questions, the editors may not print the stories because they fail to grasp the importance of demonstrations or believe that the events are incomprehensible to a wide variety of audience. Oguma (2015) believes, however, the media coverage has improved since 2011; the traditional news media now pays better attention to social protests and gives fairer reporting of them. The persistence of the movements, thus, has educated the media and cultivated their sensibility to cover demonstrations.

In addition to pressuring the traditional mass media, the antinuclear movements themselves have acted as alternative media and filled the gaps they saw. Social media and the Internet allowed the movements to become their own journalists. For example, the *Tento Hiroba* published an online daily journal and continues to do so after the removal of the tents in 2016. Those who initiated the Ōi protest published a small book to document what occurred during the 36 hours from the points of view of more than 20 participants. One of the contributors, Hasegawa (2012), noted that the local newspaper in Ōi only published criticisms of the protest as though the entire town was against the protest. However, before the protest, the tent village was visited by Ōi residents every day; they would come to chat with the activists, to bring food, and even to offer to do their laundry. They were grateful for the activists because it is difficult for them to speak against nuclear power when the town economically benefits from it. On the morning of July 2 when the protest ended, a middle-aged woman who frequented the tent village stopped by during her morning walk with her dog and told the campers to give her their dirty clothes so she could wash them. Publishing a story like this provides some balance to the media coverage even if the circulation of the book may be far smaller than a newspaper's. It not only resists being defined by the dominant narrative that tends to define protesters as disrupters but also offers the accounts of what citizenship can possibly be.

Expanded contour of protests

Finally and perhaps most intriguingly, post-Fukushima antinuclear protests expanded the contour of what a protest should look like. The large demonstrations that took place in the parks in Tokyo often included singing, drumming, dancing, and people in costumes.[5] The Ōi protest similarly incorporated these performances and more.[6] These expressions of creativity are in part due to the relative leaderlessness, a characteristic of many recent social movements around the world, including the post-Fukushima antinuclear power movements. The leaderlessness has its challenges, but it also encourages each individual to engage in meaning-making themselves and encourage accountability (Sutherland, Land, & Böhm, 2014). At the same time, these festive forms of protest drew criticisms from the public who saw the video footages of them or read about them on the Internet. A general sentiment of the criticisms is that the festival-like attitudes trivialize the movement and undermine what the movement seeks to achieve. When Tsujimura (2012a) posted her video of the Ōi protest on YouTube, many reactions included this brand of criticism. Tsujimura (2012b) had a chance to interview one of the Ōi protesters about the criticism when she saw him again at a demonstration in front of the headquarters of the KEPCO. This young man, Wataru Takasaki, who worked for an environmental NGO, responded:

> The old demonstrations were violent. You would see the protesters with firebombs and sticks. But that is an old way. *Kidoutai* is not your enemy. They are doing their job. Besides, if a movement is scary, it chases people away. And we need to stop petty competitions about whose way is the right way. We must recognize that there are many different methods to achieve the same goal. We need to understand that everyone has their strengths and what they can or cannot do in their own situation. The goal is the same. We must acknowledge the diversity within a social movement.

This idea of embracing the diversity within the movement was echoed by others. For example, one of the organizers of the Friday rally in Tokyo who goes by the name of Misao Redwolf stated,

> Expressing a sincere message is not enough. Demonstrations must appeal to young people and people with no prior experience with activism. You can do that by incorporating music, arts, and whatever expressions that suit the culture of the participants. I think there are many more people who are interested in participating in demonstrations. We must think about how we can make demonstrations attractive across generations and cultures.
>
> (TwitNoNuke, 2011, p. 52)

These words of the activists represent voices that resist a monolithic and static embodiment of protests. They push the boundaries of what protests should look like and defy policing of the boundaries that come not only from the spectators but also from within the movement.

Zooming out of Japan's antinuclear power movements and social movements, I see how these actions that resist boundaries face similar criticisms elsewhere. This past year in the United States in the wake of escalating police killings of Black Americans, Black Lives Matter protests erupted across the country. Some prominently incorporated dancing, singing, and drumming and were labeled as inappropriate (Burke, 2020). When interviewed, the performer-activists pointed out performance as a powerful form of resistance, expression of solidarity, and even the simultaneous experience of joy and pain that arts can convey (Burke, 2020). Such aesthetic and emotional expressions are fundamental human experience and express what may not be expressed through words. Performance, then, is a uniquely positioned powerful venue of political activism "as it develops new materiality, the use of bodies, and is often artistically creative, symbolic, and interactive"

(McGarry, Erhart, Eslen-Ziya, Jenzen, & Korkut, 2019, p. 19). In the end, the expanded contour of protests to embrace performativity may be one of the most significant accomplishments of the post-Fukushima antinuclear movements. As these largely leaderless movements do not appeal to specific groups but are meant for the entire citizenry (Gerbaudo, 2017), they must be inclusive of a wide range of demographics. On the subject of the criticism about the celebratory forms of protest, Oguma (2015) argues that drawing the distinction between politics and festivity is unnecessary and absurd; "Does a trumpeter who participates in a demonstration and plays the trumpet participates as an artist or for a political reason? You can't separate them. Trying to do so is nonsensical." Oguma's words remind us that diversity is not just between individuals but also within individuals. Our identity itself is multidimensional, and the connections and solidarities can rise through calling upon some of these dimensions.

Notes

1 Ten-year anniversary rallies took place across the country, although the numbers of participants were small. For example, about 450 people from 20 groups participated in a rally in Osaka City (Yamada, 2021). Similarly, a group of activists staged a sit-in in front of the headquarter of the Kyushu Electric Power Company. This group started regular sit-ins since April 2011 to stop the nuclear power stations owned by this company and had done so every day until spring 2020 when COVID-19 shut down all social activities. They now gather for the sit-in every Thursday (Yamada, 2021).

2 This was the organizers' estimate. The police believed it was about 17,000, and the news media estimated the size to be between 20,000 and 45,000 (Frackler, 2012).

3 Kan and Noda both belonged to *Minshutou* (the Democratic Party of Japan) (DPJ), a centrist political party, which became the ruling party in 2009, defeating the conservative Liberal Democratic Party (LDP) that controlled the House of Representatives since WWII. It was thus the LDP that began and grew the nuclear power program. The meeting with the MCAR may not have occurred if the LDP was in power. The DPJ dissolved in 2016 as a result of the formation of the Constitutional Democratic Party of Japan, to which both Kan and Noda belong.

4 The daily journal is found here: https://tentohiroba.tumblr.com/

5 Sociologist Yoshitaka Mouri explains that creative forms of demonstrations appeared in Japan after the U.S. invasion of Iraq in 2003 (Mouri, 2011), an event that prompted worldwide protests, including in Japan. In addition to the typical demonstration with signs and chants, artists, musicians, writers, and others in the cultural industry joined protests against the invasion and expressed their responses in a variety of ways. Mouri argues that the creativity and inclusiveness in the post-Fukushima demonstrations have roots in these earlier protests.

6 Some young men at Ōi at some point during the 36 hours of protest carried into the crowd a portable shrine on their shoulders – a kind of miniature shrine that is used in festivals. To other participants' surprise, the shrine was a giant penis. A portable shrine of this kind may sound bizarre, but several localities across Japan use it in their festivals that celebrate easy and healthy pregnancy. At Ōi, four young men carried the shrine awhile among other protesters.

References

AM Kikaku (Ed.). (2012). *Inochitachi no himei ga kikoeru [You can hear the lives scream]*. Tokyo: AM Kikaku.

Angelique, H. L., & Culley, M. R. (2014). To Fukushima with love: Lessons on long-term antinuclear citizen participation from Three Mile Island. *Journal of Community Psychology*, 42(2), 209–227. https://doi.org/10.1002/jcop.21605

Arce, M., & Rice, R. (2019). The political consequences of protest. In M. Arce & R. Rice (Eds.), *Protest and democracy* (pp. 1–21). Calgary: University of Calgary Press.

Brown, A. (2018). *Anti-nuclear protest in post-Fukushima Tokyo*. London: Routledge.

Burke, S. (2020, June 9). Dancing bodies that proclaim: Black lives matter. *The New York Times*. Retrieved from www.nytimes.com/2020/06/09/arts/dance/dancing-protests-george-floyd.html

Dominguez, G. (2015, February 25). Former Japanese PM Naoto Kan: 'Fukushima radically changed my perspective,' *Deutsche Welles*. Retrieved from www.dw.com/en/former-japanese-pm-naoto-kan-fukushima-radically-changed-my-perspective/a-18275921

Fackler, M. (2012, June 29). *In Tokyo, thousands protest the restarting of a nuclear power plant*. Retrieved from www.nytimes.com/2012/06/30/world/asia/thousands-in-tokyo-protest-the-restarting-of-a-nuclear-plant.html

France 24. (2011, July 31). *Protesters rally in Fukushima against nuclear power*. Retrieved from www.france24.com/en/20110731-protesters-fukushima-city-call-end-nuclear-energy-demonstrations-earthquake-japan

Gerbaudo, P. (2017). *The mask and the flag: Populism, citizenism, and global protest*. New York: Oxford University Press.

Hall, S. (1997). The work of representation. In S. Hall (Ed.), *Representation: Cultural representations and signifying practices* (pp. 15–64). Thousand Oaks, CA: Sage Publications, Inc.

Hasegawa, U. (2012). Okyupai Ohi no ran ga jitsugen [The Occupy Ohi revolt came true]. In *Okyupai Ohi no ran* (pp. 97–102). Saitama, Japan: Hangenpatsu Hantai Kanshi Tento.

McGarry, A., Erhart, I., Eslen-Ziya, H., Jenzen, O., & Korkut, U. (2019). Introduction: The aesthetics of global protest: Visual culture and communication. In A. McGarry, I. Erhart, H. Eslen-Ziya, O. Jenzen, & U. Korkut (Eds.), *The aesthetics of global protest: Visual culture and communication* (pp. 15–36). Amsterdam: Amsterdam University Press.

Mouri, Y. (2011). Demo no houhouron [Demonstration methodology]. In TwitNoNuke (Ed.), *Demo iko! [let's go to a demonstration!]* (pp. 31–36). Tokyo: Kawade.

Noma, Y. (2012). *Kinyo Kanteimae kogi [Friday protest in front of the Prime Minister's residence]*. Tokyo: Kawade Shobo Shinsha.

Oguma, E. (2015, September 1). Naze kantei demo wa houjirarenakatta noka? "*Shushou kantei no maede*" Shakai gakusha Oguma Eiji san [Why the demonstrations at the Prime Minister's residence were not reported? "*Shushou kantei no maede*" sociologist Eiji Oguma. *Todai Shimbun*. Retrieved from www.todaishimbun.org/oguma0902/

Okuhara, S. (2021, March 11). "Datsugenpatsu" shinto? Koizumi, Hatoyama, Kan Naoto no moto shusyo sannin, soroibumi ["No nuke" new political party?

Joint appearance of three former Prime Ministers Koizumi, Hatoyama, and Kan Naoto]. *SankeiBiz*. Retrieved from www.sankeibiz.jp/macro/news/210311/mca2103111831017-n1.htm

Saito, F. (2021, March 26). Shushokanteimae no "hangenpatsu" demo, 400kai de maku, sankashagen de shikinnan [The demonstration in front of the Prime Minister's Residence ends after its 400th event due to lack of funding accompanying the lack of participation]. *Mainichi Shimbun*. Retrieved from https://mainichi.jp/articles/20210326/k00/00m/040/459000c

Sakata, M. (2012). Ookikute, tsuyokute, yasashii kimono datta hi [When we were a large, strong, gentle beast]. In *Okyupai Ohi no ran* (pp. 69–71). Saitama, Japan: Hangenpatsu Hantai Kanshi Tento.

Sayonara-nuke.org. (n.d.). Retrieved April 3, 2021, from http://sayonara-nukes.org/2021/01/

Slodkowski, A. (2011, June 15). Japan anti-nuclear protesters rally after quake. *Reuters*. Retrieved from www.reuters.com/article/columns-us-japan-nuclear-protest/japan-anti-nuclear-protesters-rally-after-quake-idUSTRE75A0QH20110615

Sōri to hangenpatsudantai ga menkai, hanashiai heikōsen [Prime Minister and anti-nuclear power group meet but end with no common ground]. (2012, August 22). *The Asahi Shimbun*. Retrieved from www.asahi.com/special/minshu/TKY201208220622.html

Steinhoff, P. G. (2018). The uneven path of social movements in Japan. In D. Chiavacci & J. Obinger (Eds.), *Social movements and political activism in contemporary Japan: Re-emerging from invisibility* (pp. 27–50). New York: Routledge.

Sutherland, N., Land, C., & Böhm, S. (2014). Anti-leaders(hip) in social moment organizations: The case of autonomous grassroots groups. *Organization*, *21*(6), 759–781. https://doi.org/10.1177/1350508413480254

Tsujimura, M. (2012a, July 3). Ohi genpatsu saikado hantai demo [The demonstration against the restart of the Ohi Nuclear Power Plant] [Video]. *YouTube*. Retrieved from https://8bitnews.org/?p=404

Tsujimura, M. (2012b, July 11). Genpatsu saikado hantaiha no seinen ni kiku [An interview with a young man who oppose the nuclear power restart] [Video]. *YouTube*. Retrieved from www.youtube.com/watch?v=babZTsUqQ7E

TwitNoNuke. (2011). *Misao Redwolf-san*. In TwitNoNuke (Ed.), *Demo iko! [let's go to a demonstration!]* (p. 52). Tokyo: Kawade.

Usui, T. (2021, March 11). Kinyo yoru ni 10 nenkan tsuzuita "hangenpatsu" kanteimae demo no regashii towa nanica [What is the legacy of the Friday night no-nuke demonstration outside Prime Minister residence that lasted 10 years?]. *Yahoo! Japan News*. Retrieved from https://news.yahoo.co.jp/articles/21041f3c1e3459035116149e06a1cdc478ad60a8?page=1

Yamada, K. (2021, March 7). Koronaka demo "datsugenpatsu" motomete Osaka-shi de shimin dantai ga demo [Despite the pandemic, citizen groups rallied, demanding the end to nuclear]. *The Asahi Shimbun*. Retrieved from www.asahi.com/articles/ASP376QFFP37PTIL00G.html

Yoshioka, S. (2012). Okyupai Ohi no ran ga jitsugen [the Occupy Ohi revolt came true]. In *Okyupai Ohi no ran* (pp. 38–40). Saitama, Japan: Hangenpatsu Hantai Kanshi Tento.

8 Fukushima "under control"

Progress discourse and its excess

Before the disaster at the Fukushima Daiichi Nuclear Plant in March 2011, nuclear power provided about 30% of Japan's electricity, and its share was expected to grow to 50% by 2030 (World Nuclear Association, 2010). The disaster not only interrupted this plan but also led to the suspension of all nuclear reactors for the first time since 1970. Many citizens realized the danger of nuclear power for the first time and expressed their disapproval through demonstrations, social media, public polls, and more. As shown in Chapter 7, antinuclear rallies occurred following the disaster, drawing many first-time demonstrators. A number of public surveys reflected the antinuclear sentiment. When the government conducted a poll regarding Japan's future dependence on nuclear power in summer 2012, 87% of the 89,124 respondents chose the "zero nuclear" scenario out of three options: 30%, 15%, and 0% ("Pabu come," 2012). Similarly, in a telephone opinion poll conducted by *The Asahi Shimbun* in late fall 2012, a large majority preferred phasing out of nuclear power (18% "terminate immediately" and 66% "gradually phase out") whereas only 11% chose "continue" ("Genpatsu jojo ni," 2012). Reflecting this widespread opposition to nuclear power, then Prime Minister Yoshihiko Noda and his Democratic Party pledged that they would make Japan nuclear-free by 2030.

This pledge was short-lived, however. In the December 2012 election, Shinzo Abe and his neoconservative LDP regained control of the government. Abe, who ran on the ticket of lifting Japan's economy out of a recession, won by a large margin. He reinstated nuclear power as the nation's key energy and launched a prolific campaign to restore domestic and international confidence in Japan's nuclear technology. During his speech in front of the International Olympic Committee on September 7, 2013, Abe made a case for Japan to host the 2020 Summery Olympics. He assured the committee that the safety concern raised about Tokyo due to the nuclear accident in Fukushima is a "misunderstanding" and that "the situation is under control . . . the contaminated water was blocked" (Linden, 2013). His government lifted evacuation orders in early 2017 for the evacuated areas, effectively celebrating the end of the disaster. Under the new post-Fukushima

DOI: 10.4324/9781003044222-8

regulations, Japan brought nine reactors back online and are in the process of approving more.[1]

The Strategic Energy Plan (SEP) (2018), published by the Ministry of Economics, Trade, and Industry (METI), marks nuclear power's share as 20–22% by 2030. The 132-page document begins with a statement that there are points to be considered at any time energy choices are made. The very first point reads:

> First, there has consistently been no change to the position that the starting point is to take measures keeping the experience, regrets, and lessons learned of the TEPCO's Fukushima Daiichi Nuclear Power Station accident uppermost in mind. . . . [Japan] is giving the top priority to safety regarding nuclear power when realizing the 2030 energy mix and making its energy choices for 2050 and is reducing its dependency on nuclear power as much as possible as it aims to expand renewable energy.
>
> (pp. 3–4)

Although Abe stepped down as the Prime Minister in fall 2020, LDP remains in power, and Abe's successor, Yoshihide Suga, maintains the same energy policy. While no longer propagating absolute safety of nuclear power, the SEP demonstrates the Suga administration's abiding faith and confidence in preventing another Fukushima. To continue nuclear power, however, the government must regain the public's trust, which entails sincere and effective progress in the decommissioning work and the recovery of Fukushima. This requires considerable discursive labor as the public has become disillusioned by the Nuclear Village. As the material world is made meaningful through discursive representations (Hall, 1997), it is crucial for the government to communicate to the public how much strides it has made in Fukushima recovery.

The government and TEPCO and other utilities have employed a number of communicative strategies toward this end to develop the narrative of lessons learned from Fukushima in favor of continuing reliance on nuclear power. All these strategies involve visual and verbal contents to demonstrate transparency, progress, and care. The strategies particularly take full advantage of today's ocular-centric epistemology, as they heavily rely on visual rhetoric, a combination of visuality and larger textual and performative contexts that solicits a desired response from the audience (Olson, Finnegan, & Hope, 2008). Employing articulation as a theory and methodology (Slack, 2006), as I did in Chapter 4, I first discuss these strategies that shape the pronuclear, "Fukushima under control" narrative. Then, I turn my attention to the stories that challenge this narrative. I conclude the chapter with a reflection on these contrasting narratives. The texts examined here were produced by the government and the utilities and include websites, official documents (e.g., the SEP), and communication with the public in the form of media reports. They collectively construct a narrative of the goodness of the nuclear power industry through the articulation of hard work, success, and reflexivity in post-Fukushima recovery. Control over the contaminated water, progress in retrieving fuel units, and recovery of the neighboring towns feature prominently in this narrative. Although this all appears constructive, I question the end it serves. I argue that this is ultimately a destructive narrative of progress built on the logic of colonization and anthropocentrism.

Fukushima under control

Showing it like it is: closing remoteness through virtual tours

As the owner of the debilitated plant, TEPCO launched massive public relations efforts since the accident to restore its image. On its website (available both in Japanese and English), one of the main tabs on the homepage is "Fukushima eno sekinin" (Our responsibility for Fukushima). This link includes everything related to the Fukushima accident – the cause and timeline of the accident, compensation, decontamination and revitalization efforts, and decommissioning project progress. The Japanese page has a link to the information about the TEPCO Decommissioning Archive Center (DAC) that opened in November 2018 at the former location of an energy museum in Tomioka Town. Following the link, you will find a six-minute video in which a young woman, Ryoko, in a dark TEPCO uniform, gives a virtual tour of the facility. As she introduces herself and welcomes the imagined audience, the caption at the bottom of the screen gives additional information – she is from Fukushima. With soothing music in the background, Ryoko introduces each section of the two-story building in a calm, unemotional but courteous voice. The first zone is where the details of the accident are found. The visual and audio interactive tools give scientific explanations of what occurred inside the reactors. You can also see a dramatic re-creation of the control room at the time of tsunami. The second zone is dedicated to the lessons TEPCO learned from the accident. The camera shows the writing on the wall; in this section, you will hear the authentic voices of those workers who responded to the accident, and the voices will be there for future generations to hear. The third zone introduces the decommissioning work. A massive screen shows the inside of the plant as though you are there. The interactive visuals and audio guides provide information about the workers and the state-of-the-art technologies they used to control the contaminated water and remove the nuclear fuel debris. Ryoko points out that the radiation level has gone down so drastically that normal work clothes suffice in 96% of the plant property. Finally, she takes the camera to the "Communication Space" that the visitors can freely use to learn about radiation and local communities. The video ends with a screen with a message in Japanese, "We are committed to communicating in an accessible language the memories, accounts, reassessments, and lessons from the Fukushima nuclear accident and the latest information about the decommissioning work."

The DAC brochures, one in Japanese and the other in English, are also available on the Japanese webpage. It reads:

> All visitors from Fukushima Prefecture, particularly from the area around the power station, and across Japan and the world are welcome to come and learn about the unfolding of the Fukushima nuclear accident and the current status of the decommissioning work. TEPCO has a keen sense of responsibility to record the events and preserve the memory of the nuclear accident. We see it as our mission to communicate the reassessment and lessons from the accident to those both in and outside the company. We also believe it is important to demonstrate the progress we are making with decommissioning work and highlight the very long-term nature of these efforts. We will do our best to reassure the public through the revitalization underway at Fukushima; and, in cooperation with other relevant organizations, we intend to pass on the memories, reassessment and lessons from the nuclear accident to future generations.[2]

Limited numbers of visitors have the opportunity to physically visit the DAC, but the video tour makes it possible for anyone with the Internet access to virtually see the inside of the DAC and learn about the decommissioning work. At minimum, the very existence of the DAC, which is advertised through the video, communicates that TEPCO, the protagonist of the narrative of the massive decommissioning and restoration projects, is indeed committed to the work. Without information about limitations, flaws, and setbacks in the project, the virtual tours and the brochures construct an impression that the work has been categorically successful. And this communication is facilitated by a young woman from Fukushima. Ryoko, representing a young generation of Fukushima, serves as an ideal spokesperson for TEPCO. Guided by her, the introduction video serves as a trusted tool of the transparent and comprehensive communication to which TEPCO repeatedly says they are committed.[3]

Along with the DAC, TEPCO unveiled two communication tools in March 2019. According to their press release, these tools – Inside Fukushima Daiichi and Treated Water Portal Site – are introduced to

> provide detailed insight into the Fukushima Daiichi Nuclear Power station and updates on the processing of treated water stored at the facility. Through these and other initiatives, TEPCO aims to further enhance communication with external stakeholders and improve public understanding of the company's activities, including decommissioning efforts and its nuclear safety plan. . . . The company is committed to enhancing transparent dialogue with stakeholders, including local communities and news media, to address concerns and prevent misunderstandings.[4]

The same press release explains that Inside Fukushima Daiichi allows "anyone anywhere to experience what it is like to be on the decommissioning frontline." Inside Fukushima Daiichi is found under the "Visual Contents" tab where you will first encounter the message, "It is our responsibility to communicate the decommissioning work that will take 30–40 years. We will communicate what is going on at the frontline as it is." This emphasis on transparency is repeated

Within Inside Fukushima Daiichi, you first see a road with rows of buildings on the left side and rows of trees and a parking lot on the right. A narrator with soft female voice says that this is the entrance area of the plant and there used to be more than 1,000 cherry trees whose blossoms entertained many visitors in spring. She continues that, although many of these trees had to be cut down for the decommissioning project, the remaining trees are still beautiful and are loved by the workers. Then, you are taken to an interactive map with numbers on it. Clicking a number takes you to a specific section of the plant. For example, No. 7 is the area where the tanks for storing contaminated water are being constructed. The narrator explains that the workers' safety is ensured by masks and other protective gears they wear to avoid exposure to radiation. Choosing No. 4 takes you to reactors No. 2 and No. 3. The narrator speaks that the No. 3 building still shows the damage from the hydrogen explosion, but good progress is also made while worker safety is always kept in mind.

As TEPCO's press release stated, taking this tour allows anyone to *see* inside the plant as though you are really there. The company appears to tell its audience that we can witness, without a biased mediation, what it is like inside the 860-acre plant. You become the seeing agent in this tour. Yet the virtual tour is already a mediated story carefully crafted as a result of myriads of decisions: what your eyes are allowed to see, what "facts" are presented or omitted, what words and visuals are used over others and how. The introductory note about cherry trees, the repeated reference to worker safety, and emphasis on the progress that has been made, devoid of stories of struggles and setbacks, create a particular narrative of Fukushima Daiichi. Through the supposed "transparent dialogue" promoted through communication tools like these, remoteness is also supposedly removed. The invitation to experience the frontline removes physical and communication distance. TEPCO is there doing difficult work and you, as the audience, get to witness the heroic work. By showing the decommission frontline and by reporting the progress, TEPCO represents itself as authentic, ethical, responsible corporate citizen hero. TEPCO, the public, and the environment all become members of the same team. Taking the virtual tours of the plant and the DAC leaves viewers with an experiential understanding what is really going on inside Fukushima Daiichi. Beyond the tours, if you also visit other pages in TEPCO's "Our Responsibility for Fukushima," it is hard not to be impressed and overwhelmed by the details made available to the audience. You leave with an impression that TEPCO really is making great progress in the decommissioning work.

Fukushima today: promoting safe, restorative Fukushima

While TEPCO needed to build its image as a transparent, caring, trustworthy corporate citizen, the government also had to restore the public's trust in it by demonstrating its care for Fukushima. On its home page, the METI(METI) created "Fukushima Today" as one of the two main topics. Clicking the tab, you are taken to the page that includes reports of the current situation of the decommissioning work and press releases on the recovery and revitalization of the affected local communities. Fukushima Today includes newsletters introducing past and upcoming activities and events at the twelve communities that were affected by the nuclear accident in 2011. There are also short videos that introduce stories from these communities (e.g. the residents who returned, the businesses that (re)opened), interesting places to visit and delicious food to try in Fukushima, and other community revitalization news. For example, a two-minute video introduces Odaka Worker's Base, a company that launched in 2014 to rebuild the Odaka district of Minamisoma City. In the video, you meet young entrepreneurs who are excited about being part of creating a new, forward-thinking community from scratch. In another video, you meet a comedian who visits Fukushima, meets a proud young fisherman, and tries a fish dish to demonstrate the safety and tastiness of the fish caught in Fukushima. On the English-language webpage, a two-minute video, *Fukushima2020*, introduces the state of Fukushima's recovery as continuing but impressive because "Living conditions have returned to normal."[5]

In February 2021 just prior to the ten-year anniversary of the nuclear disaster, the METI launched a new storytelling project to publicize the recovery of Fukushima and began posting recovery stories on Twitter.[6] In these stories, Fukushima is branded as a prefecture with rich agriculture, high-tech industries, and resilient communities, and the decommissioning work at the debilitated plant is portrayed as impressive (e.g., state-of-the-art technologies) and economically beneficial (e.g., 4000 workers). As the purpose of Fukushima Today is to promote a safe, positive image of Fukushima, it is no surprise that there are no stories of Fukushima residents with dissenting opinions and experiences. The stories of returned residents and revival of normal activities are real and are heartwarming. However, the lack of divergent stories, some of which I discuss later, is noteworthy, as the omission helps to shape the narrative of Fukushima that, although the accident was serious, things are getting back to normal, and all will be well. It is worth asking who is served by this narrative and who is not. I will return to this question later.

The Prime Minister himself became the spokesperson of Fukushima's successful return to normalcy. In April 2019, Abe visited the Fukushima Daiichi Nuclear Power Plant and the surrounding communities. The event was widely covered by the news media, but only one visual of his visit to the plant is available for the public to see. There are slight variations of the visual depending on the news outlets (e.g., *The Asahi Shimbun, The Japan Times, Nikkei*), but they are taken at the same location and from the same angle. His face, looking toward the reactors 100 meters away, is serious, but his normal attire (a dark business suit) without any protective gear is clearly visible. A chest-high yellow plastic fence divides him from the meltdown ground zero. The fence is made of crisscross panels and low enough that you can see everything beyond the fence. This visual tells a powerful story. Abe's proximity to the crippled reactors, his business attires, and the lack of protective gear portray the normalcy of the ground zero; the plant is so safe now that even the leader of the nation can be there in his normal clothes. TEPCO's website also reports this significant event showing the same photo ("Abe shushō ni yoru," 2019).

There are other photos on their website about Abe's visit to the plant. Three photos depicted Abe expressing appreciation for the decommissioning workers: him making a speech, him handing certificates of appreciation to the workers, and a group photo to memorialize the ceremony. Abe also visited the local communities during the same trip. In the Ogawara district of Okuma Town where the evacuation order was lifted just a few days ago, Abe enjoyed eating a rice ball made with the rice produced locally ("Shushō fukkō," 2019). Abe also visited J-Village Stadium, a soccer stadium that earlier served as a hub of decommissioning work. In the *Sankei* newspaper ("Abe shushō," 2019), the main photo at the top of the article shows Abe handing a soccer ball to a boy in a soccer uniform. Behind the boy are rows of boys also in uniforms, clapping their hands to celebrate the occasion. This staged event is particularly noteworthy for what these boys symbolize. In the aftermath of the disaster, the government was heavily criticized for failing to take swift measures to protect the citizens from harmful radiation, especially children as they are more vulnerable to radiation sickness. The youths' presence at the stadium represents that the community's health has returned. Additionally, as reported widely through news media and the Prime Minister's website, this is the stadium that Abe's government chose to start Japan's leg of the 2020 Olympic torch in March 2020 to show, in his words, "the world the opening of the Recovery and Reconstruction Games and the progress of reconstruction in Fukushima" ("Visit to Fukushima Prefecture," 2019). The quote represents a particular story Abe intended to tell. As with the virtual tours created by TEPCO and Fukushima Today, the story of Abe's visit to Fukushima Daiichi and the town conveys a carefully curated positive, uplifting impression of the decommission work and recovery of the local communities, serving as an example of soft control (Aldrich, 2008). The devastated plant is mostly safe now, and the towns are being reinvigorated. Abe's appreciation ceremony at the plant and the interaction with the local youth soccer players demonstrate not only that he cares about the community and workers on the frontline but also that they are all in this recovery together: the government, TEPCO, Fukushima, and the citizens. They are all on the same team.

Impartial observers? The public as the spokespersons

The image of transparency and sincerity is promoted by yet another means. If the public is skeptical of what TEPCO and the government show and say, the best people to disseminate the image may be the ordinary audience themselves. One of such examples is found in MATCHA, a web-based tourism promotion company. According to its "About Us," MATCHA is a Japanese mega travel media company whose mission is to inform foreign visitors about the beautiful traditions and unique cultures that various localities in Japan have to offer. It seems odd to see a report on a visit to the Fukushima Daiichi Nuclear Power Plant on a travel website, especially since ordinary people are not permitted to enter the devastated plant (although there are now tours you can take to visit the neighboring towns to see the damage and restoration works).[7] The story was about the experience of two foreigners (a U.S. American businessman, Cole, and a Dutch student, Frank) living in Tokyo (MATCHA, 2019). The article begins by saying that Cole and Frank – who are referred to as reporters – decided to take a tour of the plant because they had unsettled fears about the aftermath of the accident and the future of Fukushima. To overcome the fears, they wanted to see what is currently happening at the plant. There is a small print after the introduction, stating that their visit inside the plant was made possible because it was for media coverage. The story is designed for the readers to have a vicarious experience as the two westerners tour the site.

In the first part of the story, the readers are first informed that evacuation zones are now only 2.7% of Fukushima Prefecture and that, although there are still evacuation zones near the plant, more areas are now open and the number of returnees is increasing. The radiation level, too, has decreased by 74%. This is followed by a photo of tranquil scenery of a small river with a caption informing that this area is now safe to live. The river, with abundant spring water from the nearby mountains, even welcomes salmon coming upstream in autumn. Under this photo is another photo of a road with trucks and cars going by. Cole comments, "I thought the area around the plant is deserted, but it is lively with all these utility vehicles!" The tour begins with the information that a five-hour tour of the plant accumulates only 0.04 mSv of radiation – equivalent to 2/3 of a chest X-ray. Cole and Frank visit various places inside the plant. There is, for example, a photo of them standing in front of No. 3 reactor building with a helmet and a mask over their mouth but with a light vest over their normal clothes. TEPCO's media relations spokesperson, Abe, explains that workers can now walk around like that since May of 2018 due to the removal of the debris and TEPCO's ability to control the radioactive dust. Abe guides the two young foreigners around the plant to show the tool used to retrieve melted cores, the ice walls designed to prevent groundwater flows, and the tanks storing the contaminated water.

The whole article performs a myth-buster function through the eyes of Frank and Cole. They end the first story with their final thoughts: "I was shocked that I could stand so near ground zero. The accident site was much more under control than I had expected" (Frank). "I also felt that, too. There are unsure issues like the final solution to the contaminated water problem, but I felt that the people who work there are professionals who are doing the best they can" (Cole). The story ends with the writer's note: "We fear things we cannot see. But once you know what it is, your fear will recede."

In the second part of the story, Cole and Frank explore two towns within 20 km of the plant. In Namie Town, they meet a dentist who returned to care for the community. In Minamisoma City, they meet the owner of Odaka Workers Base, the company that was also featured in Fukushima Today on the METI website. They also visit an old temple to see 1,000-year-old Buddha figures curved on the side of huge rocks and a 1,000-year-old cedar tree. The caption reads, "the 1,000 year old Buddha carvings and tree have protected people and continue to do so after the earthquake and the nuclear accident." The story ends with Cole's words: "I did not know that Fukushima is so beautiful. I was deeply touched by the natural beauty of Fukushima. It reminded me of my rural, farming hometown community."

By the end of the tours of the plant and the neighboring towns, Cole and Frank are no longer fearful about the nuclear accident. As a reader who goes on a vicarious tour with them, your fear of radiation also dissipates, and you are pleased about the progress being made at the decommission site. And it is only then at the end of the article you see in small print that their tour of the plant was sponsored by the METI in cooperation with TEPCO. When you go to the METI's Fukushima Today page, you will find links to this MATCHA article and several other stories about Fukushima attractions. However, if you find the stories on MATCHA's site, you are not given a link to the METI. If you miss the small print at the end of the article, you will not know its connection to the METI and TEPCO. Without knowing this connection, it is easy to assume that the stories of Cole and Frank are independent and impartial and are therefore trustworthy. Their foreignness adds further objectivity to their assessment of their experience. You may conclude that these brave westerners went into the melted plant to face their fears of radiation and came out with very favorable impressions of the plant on their own volition.

In discussing presence as an important concept in visual rhetoric, Hill (2004) explains presence as "an extent to which an object or concept is foremost in the consciousness of the audience members" (p. 28). Hill further argues that the strength of presence depends on the vividness of information; the actual experience is most vivid followed by moving images with sound and then photographs. A variety of texts analyzed earlier are articulatory practices to build the presence of trust through strategies of transparency. The government and TEPCO have attempted to communicate transparency by supplying vivid information, including the virtual tour, sharing of community recovery and rebuilding stories from Fukushima, and the stories of the Prime Minister and objective visitors. The visual and verbal texts hail the audience to identify with and trust them (Burke, 1969) because they are *really there*. Of course, such attempts for identification may fail, but again the pervasiveness of the discourse, especially in the absence of other articulations that challenge the discourse, shapes truth. Hence, we must turn to excavate the stories that are left out of the "Fukushima under control" narrative.

Missing stories

Inside Fukushima Daiichi

With vast public relations resources, the utility giant and the government have the ability to shape a coherent narrative of post-accident Fukushima. Yet, what has transpired in the aftermath of Fukushima is far scruffier than what the dominant narrative offers. Stuart Hall (1996) directs our attention to that which lies outside the representation; representation always has its excess – something that does not fit and thus gets left out. A good example of such excess is an experience of Madoka Kamiya, a journalist from *Tokyo Shimbun* (Kamiya, 2019). Like Cole and Frank, she was able to tour inside Fukushima Daiichi as part of Japan National Press Club. Her short report offers a very different account of the plant's cleanup progress: endless rows of enormous tanks that contain contaminated water; expansive slopes, which used to be covered with vivid azaleas and entertained local residents, now all covered by grey concrete to prevent radiation from spreading; and the daily deforestation of the woody areas of the 860-acre plant property to make room for more tanks and space for radioactive debris. Referring to the disposable masks that the workers and visitors use as an example, she comments on all the waste that is produced at the decommissioning site every day. In the story of Cole and Frank, there was a photo of them in front of No. 3 reactor. They commented on how they were impressed by the fact that they were able to stand there in normal clothes and a small mask. Kamiya's report includes a photo of No. 3 reactor as well but with a Geiger counter showing "332" in front of the reactor. The caption reads: "you can walk in your light clothes, but the radiation level here is 332μSv per hour. If you stay here for three hours, it is about 1mSv or the annual limit of radiation exposure." Kamiya's short story is by no means complete or objective (as no report is objective), but the understanding of the debilitated nuclear power plant becomes more nuanced by the inclusion of her story alongside the one by Cole and Frank.

Taming water?

The progress of the decommissioning work that TEPCO and the government so positively described also looks different when you pay attention to other reports. For example, while the virtual tour of the plant simply explains the technicalities and intended use of the "ice wall," sources other than the government or TEPCO report on the effectiveness of the wall. TEPCO informs that the ice wall is called "Land-side Impermeable Wall" and is placed around the damaged reactors to block the water flows. Super-cooled refrigerant (brine) was fed into the 100-foot-long pipes vertically placed underground to freeze the dirt around the pipes ("Land-side impermeable wall," n.d.). The frozen soil should act as a barrier to stop all water flows in and from the plant and for rainfall to bypass the contaminated soil. You are given the scientific explanation of what the wall is intended to do. This explanation serves as a presupposition that, without further information, the existence of the wall also represents the effectiveness of the wall. But there is more to the wall story. Since the nuclear meltdown, 300–500 tons of contaminated water leaked into the Pacific Ocean every day. TEPCO attempted multiple measures to stop this leakage and failed. The ice wall was attempted as the last resort. Completed in 2017, this 1.5 km long ice wall was a US$320 million project paid for by public funds. It costs about US$10 million yearly and uses 44 million kilowatt hours – enough to power 15,000 houses to maintain the wall. Despite the expensive instrument, leaks were still found in 2021 (Yamaguchi, 2021a). It has proved that the expensive wall was not very effective for the price. TEPCO reported that the wall is reducing the water flows by 95 tons a day, but whether it is thanks to the walls is being debated. Some reports show that other low-cost, low-tech methods such as pumping water from the wells around the plant are proven to be more effective than the wall ("Fukushima Ice Wall," 2018); in fact, much of the drop in the water volume is credited to the sub-drain system with some 40 wells ("High-priced," 2017).

Alongside the attempt to prevent water flows, treating contaminated water is a major daily project. The struggles and setbacks of this project are not part of the Fukushima Daiichi story told by TEPCO or the government. The earthquake and tsunami caused the foundations of the reactor buildings to crack. As much as 150 tons of groundwater seeps into the buildings daily and is mixed with radioactive materials. To prevent the contaminated water from leaking back into ground or into the ocean, TEPCO must pump up the water, store it, and treat it to reduce the radiation level of the water so it can be eventually released into the ocean. This filtering work began just a few weeks after the meltdown. The original filtering system, which used naturally occurring minerals, only filtered out cesium and not others. To solve this problem, they developed in 2013 a new system that uses artificial materials to capture 62 types of radioactive materials except for relatively benign tritium ("High-priced," 2017). Using this system, TEPCO performs a process called ion exchange (Beiser, 2018). The water is run through stainless steel tubes filled with sand grain–size particles. The particles catch the ions of radioactive isotopes such as cesium and strontium and in exchange release sodium. The treated water is stored in tanks, and the separated highly radioactive sludge is stored in sealed canisters.

This all sounds like progress, but TEPCO continues to face challenges in dealing with the contaminated water. In October 2018, they found that their filtering system did not produce the intended result; 65,000 tons of the treated water still contained radioactive materials 100 times above the safety level set by the government. In some tanks, strontium-90 – an isotope considered dangerous to human health – showed radioactivity 20,000 times above the legal limit (Sheldrick & Tsukimori, 2018). The water also contains tritium 30 times of the safety standard (Cheng, 2019). This all meant that the previously treated water stored in 1,000 tanks had to be reprocessed – a process that would likely take two years (Takenaka, 2019). In the meantime, water continues to percolate into the plant, which TEPCO must pump up and store. They must continue to build new tanks every four days (Beiser, 2018). However, TEPCO will soon run out of the space for the tanks. As the tanks' 1.37 million ton storage capacity will be full in 2022, the government announced in April 2021 the release of treated water in two years (Yamaguchi, 2021b). This news triggered protests not only from local fishermen and residents but from around Japan as well as the neighboring countries. The fishery in Fukushima is down by 85% from the pre-meltdown volume, and fish from Fukushima is sold at significantly discounted prices, as consumers are reluctant to buy fish caught in the sea near Fukushima (Takenaka, 2019).

Restored Fukushima communities?

The overwhelmingly positive representations of the recovery and revitalization of the communities around the damaged plant created by METI and TEPCO are also open for debate. The local residents who were featured in the videos are real and perhaps reflect their authentic voices, but they, too, have their excess – other local voices that are not represented in the videos. One example of this excess may be found in the experience of the tour organized by Nomado, a nonprofit organization, created by the local residents from the communities near Fukushima Daiichi to build a community through education, food, and various events. Nomado offers a free tour of the areas within 20 km from the plant so outsiders can see the affected communities. In this tour, you learn a different side of Fukushima recovery. For example, Saito (2018), who maintains a personal website, posted his experience of the tour in fall 2018. He shared a photo of a half-bulldozed mountainside to show how radiation is eliminated from mountains; you scrape off the topsoil, destroying all the trees and grasses. The contaminated surface scrapped off from mountains must be contained somewhere, but the location is undetermined. He also wrote about a dairy of farmer Yoshizawa whose cows were exposed to high dose of radiation and were condemned to be slaughtered. In the aftermath of the nuclear meltdown, the government ordered the slaughter of all livestock within 20 km of Fukushima Daiichi. Yoshizawa was one of the farmers who refused the order and continues to care for the 300 cows that no longer provide him any income.

The story of Namie Town serves as another site of contrasting stories. Saito posted photos of Namie – an abandoned school, large walled-off sections storing radioactive debris found everywhere, abandoned houses, and deserted roads with overgrown weeds. On METI's materials on Fukushima recovery, Namie is portrayed as a town where many have returned, businesses are reopening, and new exciting events are happening. Yet only 1,600 (7%) of the 21,000 residents have returned as of March 2021 (Reynold, 2021), and many former residents – 3,804 houses out of the total 7,600 – requested that their houses be simply demolished (Yamada, 2018). Losing a half of the houses is significant, and each of these houses represents a family that lost their home. These stories are, however, absent from the narrative of community recovery and revival told by the METI.

To what end?

Covering a story of a family, former residents of Namie Town, that chose to demolish their house, visual writer Toru Yamada (2018) wrote:

> Eight years after the Tohoku Earthquake and the nuclear disaster, we hear and say "recovery," but does this word really reflect the reality of the towns and families affected by the disaster? Is demolishing a house simply a demolition of a building? Is the cleared land simply a land? Are the decontaminated areas and the mountains of bags full of contaminated debris simply tainted things? Do we redeem a clean place simply by getting rid of them?

Yamada's words are worth remembering as we consume the dominant narrative of "Fukushima under control." The government launched websites to share the progress of the cleanup at the Fukushima Daiichi Nuclear Power Station and the recovery and revitalization of the neighboring towns. Near the crippled plant, TEPCO built a state-of-the-art facility to inform the public about the details of the disaster and recovery work. The company created a virtual tour for anyone to see inside the crippled plant. It also allows limited groups of people to tour inside the plant to show the progress they are making. All these, one may argue, are expected; the government and TEPCO *should* publish up-to-date information about the decommissioning work and how the communities affected by the disaster are recovering. It is their obligation to the citizens. There is no quarrel here. Nonetheless, these tours, websites, the visitor facility, and other forms of public engagement used by the state and utilities must be read as practices of soft social control (Aldrich, 2008). They are the Nuclear Village's truth-building tools that tell a limited, carefully controlled story designed to construct a narrative to foreground progress thanks to technological innovations and hard work by TEPCO and the government. It creates the appearance of closeness, transparency, and care while omitting the stories that challenge the reality they constructed. All these communicative activities contribute to the production of the regime of truth (Foucault, 1972/2010) about nuclear power in post-Fukushima Japan.

What end does this narrative serve? Recalling the SEP produced by METI, the government views the lessons from Fukushima Daiichi as the basis for their future energy policies. Their decommissioning work and Fukushima recovery are crucial part of demonstrating that they indeed learned the lessons and that they are committed to safety and the local communities. The repeated display and assertion of transparency, communication, safety, and care that show up in various verbal and visual materials they produced and widely share help to build the trust they need from the public in post-Fukushima Japan. And the trust is needed for continuing use of nuclear power. Fukushima Daiichi is no more, and no nuclear power plant will be built there. But this does not mean that there will not be another Fukushima elsewhere in the earthquake-prone island nation. With two dozens of reactors slated to be decommissioned due to their age or inability to meet the post-Fukushima safety standards ("Nihon no genshiryoku," 2021), achieving the 20–22% target means not only bringing back all feasible reactors but also likely building new reactors. The dominant narrative must be read in the context of this energy plan. "Fukushima under control" is a continuation of the pre-Fukushima hegemonic narrative that adhered to the articulation of nuclear power as safe, green, and affordable while leaving out many stories that do not fit in it (Kinefuchi, 2018). By doing so, it reproduces the regime of truth that relies on hierarchical dualism where the voice of culture (techno-science-instrumental rationality and economic growth based on nuclear power) trumps other voices (Plumwood, 1993). The concertedly positive narrative of post-Fukushima recovery and progress thus relies on the logic of colonization; it benefits the nuclear industrial complex at the expense of segments of civic society and the biosphere who are deemed disposable in the nuclear roulette.

Notes

1 This is based on May 24, 2019, information from the Agency for Natural Resources and Energy of the Ministry of Economics, Trade and Industry (METI) available at www.enecho.meti.go.jp/category/electricity_and_gas/nuclear/001/pdf/001_02_001.pdf

2 https://www4.tepco.co.jp/fukushima_hq/decommissioning_ac/pdf/leaflet-e.pdf

3 The use of young women to guide public relations is typical of corporate public relations. In fact, the public relations facilities of nuclear power plants commonly

used young women as their spokespersons. See Kinefuchi (2015) for some examples.
4 https://www7.tepco.co.jp/newsroom/reports/archives/2019/tepco-introduces-new-external-communication-tools-inside-fukushima-daiichi-and-treated-water-portal-site.html
5 www.meti.go.jp/english/earthquake/fukushima2020/index.html
6 www.meti.go.jp/earthquake/gallery/index.html
7 These tours are now categorized as part of "dark tourism" whose destinations are where tragedies occurred. www.telegraph.co.uk/travel/destinations/asia/japan/articles/fukushima-five-years-on-and-the-rise-in-dark-tourism/

References

Abe shushō, Fukushima de hisaichi shisatsu [Prime Minister Abe visits the damaged areas of Fukushima]. (2019, April 14). *Sankei*. Retrieved from www.sankei.com/photo/story/news/190414/sty1904140002-n1.html

Abe shushō ni yoru Fukushima Daiichi Genshiryoku Hatsudensho no goshisatsu [Prime Minister Abe's visit to Fukushima Daiichi nuclear power plant]. (2019, April 14). *TEPCO*. Retrieved from https://photo.tepco.co.jp/date/2019/201904-j/190414-01j.html

Aldrich, D. P. (2008). *Site fights: Divisive facilities and civic society in Japan and the West*. Ithaca, NY: Cornell University Press.

Beiser, V. (2018, April 27). Fukushima's other big problem: A million tons of radioactive water. *Wired*. Retrieved from www.wired.com/story/fukushimas-other-big-problem-a-million-tons-of-radioactive-water/

Burke, K. (1969). *A grammar of motives*. Berkeley, CA: University of California Press.

Cheng, R. (2019, March 10). *Fukushima's underground ice wall keeps nuclear radiation at bay*. Retrieved from www.cnet.com/features/fukushimas-underground-ice-wall-keeps-nuclear-radiation-at-bay/

Foucault, M. (1972/2010). *The archaeology of knowledge and the discourse on language* (A. M. Sheridan Smith, Trans.). New York: Vintage Books.

Fukushima ice wall yields limited benefits for its cost. (2018, March 11). *Nikkei Asian Review*. Retrieved from https://asia.nikkei.com/Economy/Fukushima-ice-wall-yields-limited-benefit-for-its-cost

Genpatsu jojo ni yameru 66%, Asahi Shimbunsya yoron chosa [Asahi Shimbun public survey shows 66% wants gradual phase-out]. (2012, December 3). *The Asahi Shimbun*. Retrieved from www.asahi.com/special/08003/TKY201212030003.html

Hall, S. (1996). Introduction: Why needs identity? In S. Hall & P. DuGay (Eds.), *Questions of cultural identity* (pp. 1–17). Thousand Oaks, CA: Sage Publications, Inc.

Hall, S. (1997). The work of representation. In S. Hall (Ed.), *Representation: Cultural representations and signifying practices* (pp. 15–64). Thousand Oaks, CA: Sage Publications, Inc.

High-priced Fukushima Ice wall nears completion, but effectiveness doubtful. (2017, August 16). *The Mainichi*. Retrieved from https://mainichi.jp/english/articles/20170816/p2a/00m/0na/016000c

Hill, C. A. (2004). On the psychology of rhetorical. In C. A. Hill, & M. Helmers (Eds.), *Defining visual rhetorics*. Marwah, NJ: Lawrence Erlbaum Associates, Publishers.

Kamiya, M. (2019, February 13). Jiko kara 8-nen. Fukushima Daiichi genpatsu no ima [Eight years since the accident. Today's Fukushima Daiichi nuclear power station]. *Tokyo Shimbun.* Retrieved from https://genpatsu.tokyo-np.co.jp/page/detail/932

Kinefuchi, E. (2015). Nuclear power for good: Articulations in Japan's nuclear power hegemony. *Communication, Culture & Critique, 8*(3), 448–465. https://doi.org/10.1111/cccr.12092

Kinefuchi, E. (2018). Production of the internal other in world risk society: Nuclear power, Fukushima, and the logic of colonization. In J. A. Drzewiecka, A. Jolanta, & T. K. Nakayama (Eds.), *Global dialectics in intercultural communication: Case Studies* (pp. 265–286). New York: Peter Lang.

Land-side impermeable wall. (n.d.). *TEPCO.* Retrieved May 18, 2021, from www.tepco.co.jp/en/decommision/planaction/landwardwall/index-e.html

Linden, J. (2013, September 7). Tokyo reassures IOC over Fukushima fears. *Reuters.* Retrieved from www.reuters.com/article/us-olympics-presentation-tokyo/tokyo-reassures-ioc-over-fukushima-fears-idUSBRE9860CO20130907

MATCHA. (2019, March 8). *Fukushima daiichi genpatsu wa ima? Genba wo aruite wakatta itsutsu no koto [Today's Fukushima Daiichi nuclear power plant; Five things we learned from visiting the ground zero].* Retrieved from https://matcha-jp.com/jp/6991

Nihon no genshiryoku hatsudensho no unten, kensetsu jōkyō [The operation and construction statuses of Japan's nuclear power plants]. (2021, May 3). *The ministry of economy, trade, and industry.* Retrieved from www.ene100.jp/www/wp-content/uploads/zumen/4-1-3.pdf

Olson, L. C., Finnegan, C. A., & Hope, D. S. (2008). Visual rhetoric in communication: Continuing questions and contemporary issues. In L. C. Olson, C. A. Finnegan, & D. S. Hope (Eds.), *Visual rhetoric: A reader in communication and American culture* (pp. 1–14). Los Angeles, CA: Sage Publications, Inc.

Pabu come zenshukei, genpatsu zero 87% [Public comments results: 87% supported "no-nuclear" option]. (2012, August 27). *The Asahi Shimbun.* Retrieved from www.asahi.com/politics/update/0827/TKY201208270101.html

Plumwood, V. (1993). *Feminism and the mastery of nature.* London: Routledge.

Reynold, I. (2021, March 11). Ghost towns of Fukushima remain empty after decade-long rebuild. *Bloomberg.* Retrieved from www.bloomberg.com/news/articles/2021-03-10/ghost-towns-of-fukushima-remain-empty-after-decade-long-rebuild

Saito, T. (2018, October 11). Retrieved from https://hatebu.me/entry/fukushima20km

Sheldrick, A., & Tsukimori, O. (2018, October 11). Fukushima nuclear plant owner apologizes for still-radioactive water. *Reuter.* Retrieved from www.reuters.com/article/uk-japan-disaster-nuclear-water-idUKKCN1ML162

Shushō fukkō susumu sugata sekai ni hassin, Fukushima Daiichi Genpatsu wo shisatsu [Prime Minister visits the Fukushima Daiichi nuclear power plant to show the recovery progress to the world]. (2019, April 14). *Nikkei.* Retrieved from www.nikkei.com/article/DGXMZO43725240U9A410C1I00000/

Slack, J. D. (2006). Communication as articulation. In G. J. Shepard, J. St. John, & T. Striphas (Eds.), *Communication as . . . perspectives on theory* (pp. 223–231). Thousand Oaks, CA: SAGE Publications, Inc.

Strategic Energy Plan. (2018). Retrieved from www.enecho.meti.go.jp/en/category/others/basic_plan/5th/pdf/strategic_energy_plan.pdf

Takenaka, K. (2019, March 8). *Eight years on, water woes threaten Fukushima cleanup*. Retrieved from www.reuters.com/article/us-japan-nuclear-water/eight-years-on-water-woes-threaten-fukushima-cleanup-idUSKCN1QP0MA

Visit to Fukushima Prefecture. (2019, April 14). *Prime Minister of Japan and his cabinet*. Retrieved from https://japan.kantei.go.jp/98_abe/actions/201904/_00034.html

World Nuclear Association. Nuclear Power in Japan. (2010). Retrieved from http://world-nuclear.org/info/default.aspx?id=344&terms=Japan

Yamada, T. (2018, October 16). *Machi kara ie ga kiete iku, ima Namiecho de susumu kaoku kaitai*. Retrieved from https://news.yahoo.co.jp/byline/yamadatoru/20181016-00099446/

Yamaguchi, M. (2021a, February 19). Water leaks indicate new damage at Fukushima nuclear plant. *Associate Press*. Retrieved from https://apnews.com/article/water-leaks-fukushima-new-damage-a7ecf765d0233b1cad7332ff9fed5ffe

Yamaguchi, M. (2021b, April 9). Japan to announce Fukushima water release into sea soon. *Associate Press*. Retrieved from https://apnews.com/article/japan-yoshihide-suga-pacific-ocean-16cdd288b80ef81d66315614a4e741de

9 Nuclear power, democracy, and sustainability

I began this book, arguing for a discourse-centered understanding of Japan's nuclear power. Examination of discourse reveals how our social world is created, as it is through discursive representations that meanings are created, possibilities are imagined, and actions are birthed. Throughout this book, I discussed various sites of discursive formation around nuclear power in Japan. The Nuclear Village or the nuclear-industrial complex and its supporters, including mass media, have utilized a range of discursive practices to build and disseminate a narrative that defines nuclear power as safe, indispensable, and green. They deployed various soft control strategies (Aldrich, 2008) to gain the consent of the communities where site fights occurred. These discursive and social practices all shaped the hegemony of nuclear power by creating discursive closure (Deetz, 1992). Nuclear power's righteousness was naturalized and neutralized as an objective and constant truth and was legitimized through economic, technological, and environmental rationales.

Nuclear power hegemony, however, was not without resistance even before the Fukushima nuclear disaster. It gave rise to antagonisms; threats to the well-being of communities, lack of fair citizen engagement, and the very existence of nuclear power were antagonisms that defined the limit of pronuclear discourse (Laclau & Mouffe, 1985/2001). As shown in preceding chapters, intense antinuclear movements occurred in local communities sited for nuclear facilities, and urban activists labored to educate the public and served as citizen watchdogs of the government and the utilities. While not recognized widely, these pre-Fukushima movements helped to prevent many reactors from being built and applied pressure to the Nuclear Village for accountability and transparency.

The Fukushima nuclear disaster tore the fabric of nuclear power hegemony that had survived earlier threats such as Chernobyl and Tokai-mura accidents. Following Fukushima, widespread and diverse antinuclear movements occurred throughout Japan, calling for the end to the nuclear power program. These movements, although considerably curtailed in the recent years, have been hailed as "new" social movements in Japan as they have

DOI: 10.4324/9781003044222-9

mobilized a large number of non-activist citizens and were diverse in expressions and strategies. Although Japan has a history of mass protests, the post-Fukushima antinuclear movements made protests accessible, ordinary, and diverse in expressions.

Notwithstanding the predominant call for the end of nuclear power, the state resumed the nuclear power program in July 2012. To rebuild public trust, the Nuclear Village launched extensive public relations and media campaigns to showcase the progress in the decommissioning work and the revitalization of Fukushima, while leaving out problem stories that do not fit in the progress narrative. These campaigns have served as a form of soft control to solicit public approval of the continuing existence of the Nuclear Village. The Fukushima nuclear disaster may have forever changed the public's view of nuclear power, and citizens may not be easily persuaded by the pronuclear discourse. To overcome this obstacle, the nuclear-industrial complex engaged in wide-ranging communicative strategies to build trustworthiness, scientific expertise, and community care. Hence, the struggle over the regime of truth about nuclear power continues.

What do all these analyses of discourses tell us? My research began with a naïve and simple curiosity about discourses around Japan's nuclear power. As I began to look closely into various aspects of the nuclear discourses, however, two questions kept surfacing for me, sometimes conspicuously and other times quietly. By way of concluding this book, I engage these two questions – a question of democracy and a question of sustainability.

Civic voice and discursive equity deficits in Japan's democracy

Japan's contemporary democracy was established in 1947 under the Allied occupation. After losing WWII, a western-style representative democracy was instituted as part of the new constitution drafted under the supervision of Douglas MacArthur and accepted by Japan's bureaucratic leaders. Because of this "top-down democracy" and because Japan was one of the first non-western, non-white, non-Christin countries to adopt democracy, Japan has been seen as a successful example of how non-western countries could adopt democracy in a relatively short period of time (Haddad, 2012; Otmazgin, Galanti, & Levkowitz, 2014). Still, there have been many criticisms that indicate systematic flaws that undermine democracy. Kobayashi (2011), for example, argues that Japan's democracy is malfunctioning because there are gaps: the members of parliament not pursuing the policies that reflect the public will, citizens' voting decisions not reflecting their assessment of the candidates' platforms, and politicians privileging their constituency and their own career over the good of the nation. Smith (2018) examined Japan's "dynastic politics" that grants unearned privileges to aspiring politicians who come from a pedigree of politicians – a practice that is not uncommon in other democracies, but more pervasive in Japan. Otmazgin et al. (2014) argue that, despite many criticisms that suggest Japan's failure to act democratically (e.g., non-transparent decision-making practices, institutional corruption, growing social inequalities resulting from faulty policies), Japan's democracy is steadfast and is in the rank of a "maturing democracy" that is capable of reflexivity regarding its own flaws and of self-correction. Haddad (2012) is also of the opinion that "Japanese democracy is 'real' because it is fully democratic; it is a government of, by, and for its people" (p. 27).

Yet, at least within the context of my study, the evidence points to weaknesses in Japan's supposed democracy. The veteran activists' experiences, the lack of state response to the antinuclear movements, and the vast webs of discursive control by mega corporations and the state suggest considerable shortfalls of real influence citizens have on shaping national policies. Almost four decades ago, Benjamin Barber (1984/2003) warned against liberal democracy as "thin" democracy and called for a "strong" democracy that is grounded in direct participation. He saw representative democracy that embodies liberalism as a shallow form of democracy because it denies individuals the exercise of the responsibility for their values, beliefs, and actions – a necessary condition for freedom, equality, and social justice. By allowing and even encouraging the independent grounds by the representatives, this democracy sabotages the very will of citizens it is supposed to uphold. For Barber, this form of democracy is neither democratic nor even political. Instead, he proposed that what we need is "strong democracy" that "resolves conflict . . . through a participatory process of ongoing, proximate self-legislation and the creation of a political community capable of transforming dependent private individuals into free citizens and partial and private interest into public goods" (p. 151). Participatory democracy makes public deliberation available for all who have a stake on the matter (Benhabib, 1996) and places public reasoning among the participants who are considered equal (Cohen, 1996). One may argue that direct participation is impossible to attain in Japan, a country with 126 million people. However, for a nation-state to exist as a *democratic community*, the principle of participation and civic voice that participatory democracy insists must have a place in influencing national policies. The struggles that local and urban activists experienced and the stories that are unheeded signal the dire lack of opportunities to make their voices heard. These voices challenge us to consider what democracy should look like.

One interesting aspect of Japan's nuclear site fights with regard to democracy is the local communities' ability to block the construction of nuclear facilities as detailed in Chapter 5. But, as we saw, the financial incentives were and are a powerful tool of persuasion to which many economically struggling rural communities submit. In our interview, Mr. Koyama of Mihama-no-kai told me that the problem of nuclear power cannot be simply reduced to pronuclear versus antinuclear precisely because of financial struggles that plague rural communities:

> If I give you an example, when the Ōi Nuclear Power Plant was going to be built, there was a strong opposition. But that opposition was suppressed when the government promised to build a bridge that the town desperately needed. Another example is Rokkasho-mura. The village suffered from depopulation and lack of work. During winter, men had to become migrant workers in far-away cities to earn a living. If the reprocessing plant was built, it would create employment. So, the village agreed to host the controversial facility.

These examples ring true for virtually all communities where nuclear facilities were built. Thus, while it is true that local communities have the power to reject nuclear facilities, the communities and power companies are never equals in site fights. By design, genuine citizen participation in decision-making was virtually made unattainable before fights even began.

And what about the voices of the citizens outside these ground-zero communities? Energy matters have long been thought to be the domain of the state and experts (Kinsella, 2015) in which citizens lack legitimate decision-making voice and participate only as consumers. The antinuclear movements have shown that they are no longer satisfied with that role. Should citizens have a say in deciding their society's energy future? My study leads me to say, "Yes, they absolutely should." Ten years after the nuclear accident, a majority of Japanese want to see nuclear power phased out. In a national opinion poll conducted by Japan's public television, NHK, 50% answered "must reduce" and 17% chose "close all" in response to the question about the future of nuclear power ("Kokunai no genpatsu," 2021). Similarly, the national poll conducted by Japan Association for Public Opinion Research showed that 8% of the citizens believe that Japan must end the nuclear power program immediately, and 68% answered that the program must end sometime in the future ("Datsu genpatsu," 2021, March 7). As discussed in Chapter 7, more than 8.8 million people signed the campaign to end Japan's nuclear power as of March 3, 2021. The public's wish is consistent across different surveys. However, Japan currently lacks a political mechanism to have these voices counted. In Switzerland, where a national referendum system exists, a referendum to phase out nuclear power and to develop renewable energy passed with more than 58% of the votes in 2017 despite a pronuclear campaign that urged the citizens to vote "no" ("Switzerland votes," 2017). Referendums' real value may come from public dialogues they prompt; a study of social media debates leading up to the Swiss referendum showed that multiple, diverse communities with conflicting positions engage in debates with each other (Arlt, Rauchfleisch, & Schäfer, 2019). One may argue that the small population of Switzerland makes this type of national citizen participation possible. Referendums as a tool of democracy are not without problems, either, as they could be vulnerable to political corruption and lead to ill-informed decisions. But referendums are possible and can be a just tool when there is a clear gap between the public will and policy as in the case of Japan's nuclear power. In fact, referendums have been proposed in at least six prefectural governments after the Fukushima disaster with regard to the restart of nuclear power. The citizens collected signatures that represent at least one-fiftieth of the eligible voters that are required to bring a referendum proposal to the governor who then decides whether to support the proposal and send it to the prefectural senate. All six proposals have been denied either by the governor or by the senate. Prefectural senates claim that they were already elected by citizens and thus their opinions represent their constituents, and some governors argue that nuclear power matters must be decided by the national government because it is a national policy ("Genpatsu jūmin tōhyō," 2020). These conclusions end the discourse by disqualifying public voice (Deetz, 1992). Thus, even when citizens exercise their voice, the current system does not allow their will to be counted. This is a grave flaw that undermines civic society that is fundamental to democracy, representative, or otherwise.

Referendum is one tool of participatory democracy. There may be others. Citizens' voices can be, for example, heard through increasing the ability for them to choose energy sources. Regardless of the methods of citizen participation in democratizing energy, citizens need to be given the opportunities to be informed about the matter at hand fairly and fully. This is why attention to discourse is crucial. If citizens are given only a limited, partial discourse of the matter, their participation cannot be fair. We saw in Chapter 8 that the state and TEPCO stressed transparency in their post-Fukushima public relations and media campaigns to rebuild public trust. Transparency in principle is righteous, but it is not genuine transparency if the discourse is curated in favor of one-sided representation. A transparent representation discloses the ways subsidies and other financial incentives (which come from taxes) are used. A transparent representation includes and engages diverse and conflicting voices.

Regrettably, I continue to see examples to the contrary. A latest example is the Great East Japan Earthquake and Nuclear Disaster Memorial Museum (Denshōkan) that opened in September 2020 in Futaba Town, about 2.5 miles west of the devastated plant. This US$50 million prefectural facility, built with a government grant, aims to educate visitors about the compound disaster and its impacts and to pass on the memories and learnings to future generations as lessons in disaster prevention and mitigation. Within four months of opening, 34,000 people visited the museum despite the pandemic restrictions. Like TEPCO's decommissioning center and Cole and Frank's tour in Chapter 8, Denshōkan is designed for the visitors to experience the disaster through images, stories, and optional tours to damaged buildings. Another optional attraction is storytelling by area residents so visitors can hear "the real voices of people who experienced the disaster and deepen your understanding about disaster prevention through feeling the experience as if it were your own" ("General Training," n.d.). Nothing may be more powerful than hearing the stories from the people who have direct experience, and this sounds like an effort to include marginalized voices. Indeed, one of the storytellers interviewed by journalist Shōhei Makiuchi (2021) commented that sharing his story is his way of both remembering and letting out the pent-up anger and vexation.

As tragedies and disasters become distant past, memorial museums play a crucial role in keeping the experiences of the past alive. But at Denshōkan, the storytellers are not entirely free to share their thoughts. The manual they are required to follow prohibits them from criticizing specific groups, including the state and TEPCO, and museum employees check the stories beforehand and may even edit the stories (Yamashita, 2021). Visitors' questions are also monitored by museum employees assigned to the storytelling section (Watanabe, 2020). These practices carefully control the direct experience discourse.

Visitors are, however, not naïve consumers of the information. In a citizen dialogue event organized by a high school civics teacher, 23 visitors, who independently visited Denshōkan, shared their impressions. Many felt that the museum stressed recovery and lessons framed in terms of better preparation for future disasters and evacuation and was missing contextual information such as how nuclear power came to the region in the first place and lacked the recognition of human errors as a critical factor in the nuclear disaster (Makiuchi, 2021; Watanabe, 2020). Makiuchi (2021) concluded that citizens have the responsibility to raise voice to improve Denshōkan.

Makiuchi's conclusion serves as a moment of critical reflection on nuclear power discourse in democracy. Citizens do have the responsibility to use their voice to call out unfair or unjust practices by their government. We saw many examples of these citizens in this book. However, the governments, national and local, must have the capacity to change in response to citizens' voices, which has been lacking in the bureaucratic responses to citizens. As Laclau (2001) argues, hegemony is constitutive of democracy as particular forces must occupy the place of power; however, this place must stay empty, and the particularity of the forces that occupy the place must never assume universality. Democracy requires that the gap between particularity and universality remains and reproduced. Hence, power and discursive struggles are constitutive of working democracy, and different forces must be able to participate in the struggles. This means that, on the one hand, citizens must become critical consumers of the dominant narrative and challenge the narrative with alternative stories to build legitimacy. On the other hand, a state that claims to be a democracy must remove discursive closure and build the capacity to respond genuinely to the stories and change accordingly. This may be one of the most important lessons from Fukushima and the last ten years that Japan's democracy must heed.

Toward a sustainable world

This project led me to also consider a question of sustainability not only in the context of Japan but also as a question exigent to the global community. Sustainability is commonly understood as involving the continuing existence of economy, society, and the environment. This triple bottom line has been used by the Nuclear Village to advance their discourse. The Japanese government eagerly supported the development of the nuclear power program on the basis of Japan's lack of natural resources for energy production that is critical for economic activities and social world as a whole. The environmental reasons were later added when global warming due to carbon emissions became a global concern. In the wake of the Fukushima nuclear disaster, the world stopped to ponder the risk of nuclear power as a real consequence of living in a world risk society (Beck, 2009). There were worldwide protests against nuclear power, and momentarily nuclear power became incompatible with the idea of sustainability. Yet that moment appears to have passed. The United Nations' Intergovernmental Panel on Climate Change sees nuclear power as a necessary energy for deep decarbonization that is necessary today. When the United Nations announced the 17 Sustainable Development Goals (SDG) in 2015, the nuclear power industry promptly positioned itself as a core player in attaining the goals. The International Atomic Energy Agency (2016) published *Nuclear Power and Sustainable Development* to address nuclear power's central role in SDG 7: "Ensuring access to affordable, reliable, sustainable and modern energy for all." The document framed SDG 7 as the goal foundational to other SDGs, encompassing climate action, poverty, hunger, well-being, and economic growth. To this, World Nuclear Association (2021) added that nuclear energy is vital in fighting the pandemic because it can supply electricity reliably. These texts, thus, reproduce the themes of potency and entelechy (Kinsella, 2015), positioning nuclear power as *the* effective vehicle to build a sustainable world.

Currently 31 countries use nuclear power, and some 30 more countries without nuclear power are considering or are already in the process of starting a nuclear program (World Nuclear Association, 2021). Nuclear power enjoys strong support not only from the political leaders of these nations, right and left, but also from some strands of environmentalists. James Lovelock, a world-renowned scientist and environmentalist known for the Gaia Hypothesis, views nuclear power as the safest, greenest, inexpensive gift to humans (Moss, 2014), and Michael Shellenberger, the founder of Environmental Progress, maintains that "only nuclear can lift all humans out of poverty while saving the natural environment" (Bailey, 2017). If not pronuclear, many environmentalists are in the grey area or take "the lesser of the evils" approach when it comes to nuclear power because of the threat of climate change (Harris, 2013; Plumer, Fountain, & Albeck-Ripka, 2018). Thus, today in the backdrop of accelerating climate change combined with the increasing demand for electricity, nuclear power is dubbed an essential energy to build a sustainable world. This is the articulation I question to conclude this book.

Too often nuclear power is debated in terms of its safety and carbon emissions during its operation. As we saw in this book, Japan's competing discourses have essentially focused on this frame as well. However, the frame leaves out many problems that accompany nuclear power. The assessment of nuclear power's sustainability, I argue, must take a lifecycle approach and the notion of well-being that is inclusive of what Val Plumwood (2008) calls "shadow places." These are places that provide us with the material and ecological support but are outside our consciousness and evade our responsibility, because they are distant from us. But as we never exist in isolation, "the atomism and hyper-separation of self-sufficiency is never a good basic assumption, for individuals or for communities. Communities should always be imagined as in relationship to others, particularly downstream communities, rather than as singular and self-sufficient" (Plumwood, 2008, p. 139). Barry Commoner's (1974) four ecological principles give support to Plumwood. These principles assert that everything is connected to everything else, everything must go somewhere (there is no "away"), nature knows best (i.e., excessive human intervention is detrimental to the natural system), and there is no such thing as a free lunch (if you take something, you must pay for it). The notion of shadow places and the ecological principles, along with the logic of colonization (Plumwood, 1993, 2002), provides a much-needed framework for reflecting on nuclear power's sustainability and its claimed contribution to a sustainable world.

Many shadow places exist in nuclear power. Mining and milling of uranium are one, but this issue is rarely discussed in neither pronuclear nor antinuclear discourse. The IAEA's (2016) *Nuclear Power and Sustainable Development* frames uranium as a matter of resource availability and cost-effectiveness; uranium is abundant and affordable and thus makes nuclear power sustainable. This argument makes no mention of environmental, health, and cultural costs that accompany mining. Uranium mining necessarily contaminates a vast area with chemicals and radioactive rocks and pollutes increasingly scarce groundwaters (Natural Resource Defense Council/NRDC, 2012). A used-up mine and its surrounding area become a superfund site. In the United States alone, there are 160,000 abandoned uranium mines (Morales, 2016), some 4,000 of which still wait for cleanup, and it is the taxpayers and residents downstream who bear the cost (NRDC, 2012). Contamination of the environment is inextricably related to environmental justice as mining undermines the public health of the host communities, most of which already struggle with poverty and scarce economic opportunities. This is a problem for any community but one that disproportionately affects indigenous peoples, because nearly 70% of the world's uranium is underneath indigenous peoples' lands in Africa, Asia, Australia, and North and South America (Endres, 2009). The lasting impact of uranium mining on the Navajo Nation serves as a sobering lesson. Between 1944 and 1986, mining companies extracted 4 million tons of uranium from more than 500 mines on the Navajo land. Mining ended with the end of the Cold War, but the workers and communities suffered and continue to suffer from deaths and illnesses from uranium poisoning (NRDC, 2012). Even today, high levels of uranium are found in the urine of Navajo residents, and uranium is found even in babies born today (Morales, 2016). Drinking water is chronically contaminated by uranium exposure (Malin, 2015). These problems exist at all uranium mining communities. In Australia and Canada, major suppliers of uranium for Japan, the Aboriginal and First Nation people have been dealing with increased cancer cases, contaminated water, and pollution of their sacred sites (Bryant, 2014; Keeling & Sandlos, 2009). As Commoner (1974) argued, there is no free lunch in ecology; what are taken from the Earth comes with price, and it is these mining communities that are paying the price.

Another shadow place concerns nuclear waste. Like mining, this issue is given only a fleeting attention in a common discourse of nuclear power's environmental and economic benefits. Nuclear power proponents deem the waste issue benign. The World Nuclear Association (WNA) (2017), for example, seeks to discredit ten pervasive "myths" of nuclear waste. Take the fourth myth: "Nuclear waste is hazardous for tens of thousands of years. This clearly is unprecedented and poses a huge threat to our future generations." The WNA refutes this: many industries produce toxic waste; nuclear waste "naturally decays" and has "a finite radiotoxic lifetime"; only a small part of nuclear waste is highly radioactive; and hazardous waste is within the dose specified by the "international conventions." Throughout their responses, the WNA constructs the toxicity of nuclear waste as innocuous by drawing upon familiar logics such as naturalness, assumed legitimacy of international conventions, and comparison to something supposedly worse (regardless of whether such a comparison has merit). This construction contributes to negation of nuclearity that Kinsella (2019) observes; the material removal of nuclear waste from the production site can also symbolically remove nuclearity of the waste. Notwithstanding this discursive de-nuclearization, discarded isotopes remain radioactive for many decades (strontium-90 and cesium-137), and even 24,000 years in the case of plutonium-239 (United States Nuclear Regulatory Commission, n.d.). And they must be stored somewhere, as there is no such thing as "away" (Commoner, 1974). This "somewhere" will inevitably be in or near communities that are already structurally (racially, socioeconomically, and culturally) marginalized. The case of the proposed high-level waste site at Yucca Mountain, a land of the Navajo Nation, demonstrates this environmental injustice that some call "nuclear colonialism" (Endres, 2009). In Japan, a small struggling fishing town, Suttsu-cho, in Hokkaido is being considered for the same purpose. However, to date, no country has found an acceptable way to manage high-level wastes, making nuclear power a uniquely irresponsible technology.

Economically, too, nuclear power creates shadow places. We saw in Chapter 5 that economic benefits are used to persuade communities to accept a nuclear power plant in their backyards. However, examples from Japan reveal that nuclear power brought immediate and temporary financial wealth at the cost of the long-term well-being of the communities (Kinefuchi, 2018). As local economies grow to support the nuclear power plant, land-based economy becomes debilitated. This creates serious economic and environmental sustainability challenges in the long term, because nuclear power reactors must eventually be decommissioned, and the towns are left then without resilient locally grounded economies for future generations (Kinefuchi, 2018). This is why the citizens in towns such as Maki-machi and Hidaka-cho discussed in Chapter 5 fought to protect their communities. The scenario of economic and environmental challenges is not unique to Japan, however. The Pew Charitable Trusts reported that the closure of nuclear power plants in the United States causes fiscal hardships to local communities whose economy heavily depends on the plants in the form of tax payments (Moore, 2018). Parallel stories of economic and environmental hardships are found in uranium mining towns that support the uranium industry's expansion even as they struggle with the irreparable damages done to their environment and health because they are constrained by the legacy of mining and resultant environmental and health deteriorations, persistent poverty, and spatial isolation (Malin, 2015).

The nuclear power program as a whole has become financially exorbitant. Haas, Thomas, and Ajanovic (2019) conducted a historical analysis of nuclear power construction costs and concluded that the nuclear power was never really "cheap" because the construction costs and the construction time were systematically and vastly underestimated. Additionally, the costs rose due to additional safety requirements, need for better materials, and rising labor costs. This unaffordability is one of the reasons why two reactors under construction in South Carolina were abandoned in 2017 after $9 billion invested and more than 60% completed, because it could cost as much as $25 billion or more than the twice of the initial estimate (Plumer, 2017). Nuclear waste, too, is immensely expensive. U.S. taxpayers are paying utilities at least a half billion yearly just to maintain the waste from nuclear energy, because there is no geological repository (Feldman, 2018). One nuclear plant, the Main Yankee, that ended its operation in 1996 still costs taxpayers $35 million per year (Wade, 2019). These are financial costs that accompany nuclear power even without accidents and will be the burdens that inevitably rest on the future generations. Shadow places then exist in temporal terms as well.

When we articulate nuclear power, considering its lifecycle and its extensive impacts environmentally, socioculturally, and economically, it becomes clear that nuclear power relies on the logic of colonization that uses spatial, communicative, temporal, and technological remoteness and erasure of "shadow places" (Plumwood, 1993, 2002, 2008). Most nuclear energy consumers are spatially removed from the shadow places – the experience of mining and milling (the vast majority of which are in indigenous lands), nuclear power plants, and nuclear wastes, making it difficult for them to see the destructions of the lands, communities, health, and the environment that result from these activities. The spatial remoteness is closely tied to technological remoteness; most of us who benefit from nuclear technology as consumers are far removed from the places that become wastelands because of the very technology. The authorities who make decisions related to nuclear power are removed from the impacts of the decisions they make, and the decisions made elsewhere systematically fall on the communities (including the ecosystems and non-human animals) around the nuclear sites.

Nuclear power takes advantage of temporal remoteness as well. This is most apparent in the case of accidents and nuclear waste. The area near Chernobyl is deemed inhabitable and will be so for the next 24,000 years, and the ongoing cleanup has already cost over US$235 billion according to the Belarus Foreign Ministry (2009). Fukushima cleanup and decommissioning work is likely to last the next 40 years and is estimated to cost as much as US$740 billion (Japan Center for Economic Research, 2019). Nuclear waste, too, leaves substantial and perpetual burdens on the future generations as it needs to be managed for tens of thousands of years into the future. As Kinsella (2019) observes, the assumptions that undergird future manageability of nuclear waste are fundamentally flawed because distant future can bring many unfathomable changes; when the literate civilization is only about 5,000 years old and when we do not know if humanity will exist 10,000 years from now, how can we be adequately responsible for our waste? The future generations, humans, and more-than-humans, then, are also shadow places that are systematically erased from our consciousness when nuclear power is dubbed sustainable. Although Japan's antinuclear discourse has tended to focus on immediate geopolitical spaces and present impacts, what makes the hegemonic, pronuclear articulation of nuclear power as sustainable energy troubling is the far-reaching consequences of nuclear power's materiality. The enormous social, ecological, and economic costs across time and space make nuclear power adversative to a sustainable world.

There is extensive evidence that decarbonization is attainable without nuclear power. Iceland and Paraguay are already 100% renewable, followed by Costa Rica (99%), Norway (98.5%), Austria (80%), Brazil (75%), and Denmark (69.5%), and large regions such as Germany's Mecklenburg-Vorpommern and New Zealand's South Island, Canada's British Columbia and Quebec have already achieved 100% renewables or close to it (Brown et al., 2018). Powering the planet with renewable energy is not only possible but affordable (Brown et al., 2018; Lovins, 2011; Chestney, 2021, June 22;

Tierney & Bird, 2020). As we transition to renewable energy, however, we need to seriously question where we are heading. If a renewable energy world only means replacing fossil fuels and nuclear power with solar, wind, and others without reducing our energy consumption, it will not create a sustainable world. Hence, it is not enough to argue about unsustainability of nuclear power, but we must raise the question of a sustainable world. One of my interviewees, Mr. Nishio, noted that, if a society is serious about global warming, it needs to first reduce the use of electricity, and I agree. Our modern industrialized culture takes unending economic growth as a taken-for-granted principle of life, which requires incessant use of electricity. But nothing can endlessly grow. Not on a finite planet. Sustainability must, first and foremost, mean the sustainability of the planet. Some 30 years ago, economist Harman Daly (1990) observed that we use economic growth and wealth synonymously, but beyond the optimal scale in relation to the biosphere, we actually become poorer. This is so because our excessive exploitation of the natural world threatens our very lifestyle and eventually life. One framework useful for understanding the optimal scale is the nine planetary systems and their thresholds identified by Johan Rockström and his colleagues (2009) of the Stockholm Resilience Center (SRC), popularly known as nine planetary boundaries. A 2015 study (Steffen et al., 2015) from the SRC determined that, due to human activities, four of the nine systems – climate change, loss of biosphere integrity, land-system change, and altered biogeochemical cycles – are no longer in the "safe" zone.

These are real antagonisms we face today as a global community. The question is, what do we do about them? In advancing his theory of reflexive modernization, Beck (2006) argued that risks that accompany late modernization have "involuntary enlightenment" and "enforced cosmopolitanism" functions because they drive people and societies to talk to each other across differences about the risks in order to device ways to deal with them. Indeed, we live in world risk society with ecological crises that do not respect human-made borders and differences. To an extent, international and intergroup conversations and initiatives to mitigate the risks are happening everywhere, as shown, for example, in the United Nations' conventions on climate change and on biodiversity. However, these efforts often do not produce genuine cultural changes fast enough if at all. This is so because economic growth and accompanying insatiable consumption of energy are the law of the land and are pursued competitively by individual states for the benefits of their own at the expense of human and ecological shadow places.

We need a much deeper and wide-ranging intentional discursive work at every level of society. A framework like the planetary boundaries does not exist simply to record the deteriorating earth systems. It was created to inspire discourse, dialogue, actions, and solutions, and that has to happen in schools, in communities, on the legislative floors, and in national policies. Toward that end, if we take sustainability seriously at this juncture of multiple environmental, climate, and social equity crises, we need an eco-justice lens to guide our activities in every aspect of our lifeworld, including our approaches to the problematic of energy. Ecojustice is inclusive of environmental and social justice, but the scope of justice extends beyond anthropocentric concerns. We cannot achieve a sustainable world solely for humans. Nor can we protect the environment without caring for the well-being of human communities. This is so because the crises we face today exist at the intersection of social and environmental. Or it may be more accurate to say that social is and has always been part of what we call the environment. There are many rich insights that can guide our discourse: kinship and sacred relationship between humans and nature (e.g., Armstrong, 2008; Sanchez, 1994), the concept of *buen vivir* from South America (Chassagne, 2020), earth democracy (Shiva, 2005), ecological principles (Commoner, 1974), communion of subjects (Berry, 1988), meaningful and responsible multiple relationships to place (Plumwood, 2008), to just name a few. All these visions from diverse philosophical, cultural, and disciplinary traditions are grounded in the most primary fact of the ecological systems – interconnectedness. We are interrelated with everything else across the spatial and temporal dimensions, which creates intrinsic responsibilities for that everything else. Identity, society, technology, economics, education, and all other aspects of our lifeworld are shaped in this web of relatedness. This is not a utopian view of the world that is nice to have but is the ecological fact that must be heeded by the world risk society. In this ecojustice orientation to the world, there will be still power struggles, as they are a fact of diverse world. However, ordinary citizens, non-human species, ecosystems, and future generations have genuine voice in the struggles. It is with this understanding that nuclear power, energy future, and what it means to create a sustainable world must be discoursed.

References

Aldrich, D. P. (2008). *Site fights: Divisive facilities and civil society in Japan and the West*. Ithaca, NY: Cornell University Press.

Arlt, D., Rauchfleisch, A., & Schäfer, M. S. (2019). Between fragmentation and dialogue. Twitter communities and political debate about the Swiss "nuclear withdrawal initiative." *Environmental Communication*, 13(4), 440–456. https://doi.org/10.1080/17524032.2018.1430600

Armstrong, J. (2008). An Okanagan worldview of society. In M. K. Nelson (Ed.), *Original instructions: Indigenous teachings for a sustainable future* (pp. 66–74). Rochester, VT: Bear & Company.

Bailey, R. (2017). The rise of atomic humanism. *Reason*. Retrieved from https://reason.com/2017/09/10/the-rise-of-atomic-humanism

Barber, B. R. (1984). *Strong democracy: Participatory politics for a new age*. Berkeley, CA: University of California Press.

Beck, U. (2006). Living in the world risk society. *Economy and Society*, 35(3), 329–345.

Beck, U. (2009). *World at risk*. Cambridge: Polity Press.

Belarus Foreign Ministry. (2009). *Chernobyl disaster*. Retrieved from https://chernobyl.undp.org/russian/docs/belarus_23_anniversary.pdf

Benhabib, S. (1996). Toward a deliberative model of democratic legitimacy. In S. Benhabib (Ed.), *Democracy and difference: Contesting the boundaries of the political* (pp. 67–94). Princeton, NJ: Princeton University Press.

Berry, T. (1988). *The dream of the earth*. Sierra San Francisco: Club.

Brown, T. W., Bischof-Niemz, T., Blok, K., Breyer, C., Lund, H., & Mathiesen, B. V. (2018). Response to "Burden of proof: A comprehensive review of the feasibility of 100% renewable-electricity systems." *Renewable and Sustainable Energy Reviews, 92*, 834–847. https://doi.org/10.1016/j.rser.2018.04.113

Bryant, K. (n.d.). Uranium mining, waste and Indigenous Australia. *Overland Literary Journal*. Retrieved June 7, 2021, from https://overland.org.au/2014/09/uranium-mining-waste-and-indigenous-australia/

Chassagne, N. (2020). *Buen Vivir as an alternative to sustainable development: Lessons from Ecuador*. London: Routledge. https://doi.org/10.4324/9781003023074

Chestney, N. (2021, June 22). Falling renewables costs undercut new and some existing coal plants – study. *Reuters*. Retrieved from https://www.reuters.com/business/energy/falling-renewables-costs-undercut-new-some-existing-coal-plants-study-2021-06-22/

Cohen, J. (1996). Procedure and substance in deliberative democracy. In S. Benhabib (Ed.), *Democracy and difference: Contesting the boundaries of the political* (pp. 95–119). Princeton, NJ: Princeton University Press.

Commoner, B. (1974). *The closing circle: Nature, man & technology*. New York: Bantam Books.

Daly, H. E. (1990). Toward some operational principles of sustainable development. *Ecological Economics, 2*, 1–6. https://doi.org/10.1016/0921-8009(90)90010-R

Datsu genpatsu 76% kenen nezuyoku Fukushima genpatsu jikoga eikyo, Zenkoku yūsō yoron chōsa [The national study shows 76% of the public calls for no-nuke, showing strong influence of the Fukushima nuclear accident]. (2021, March 7). *Okinawa Times*. Retrieved from *https://www.okinawatimes.co.jp/articles/-/717608*

Deetz, S. A. (1992). *Democracy in an age of corporate colonization*. New York: State University of New York Press.

Endres, D. (2009). The rhetoric of nuclear colonialism: Rhetorical exclusion of American Indian arguments in the Yucca Mountain nuclear waste siting decision. *Communication and Critical/Cultural Studies, 6*(1), 39–60. https://doi.org/10.1080/14791420802632103

Feldman, N. (2018, July 3). The steep costs of nuclear waste. *Stanford Earth Matters*. Retrieved from https://earth.stanford.edu/news/steep-costs-nuclear-waste-us#gs.3998ot

General Training. (n.d.). *The great east Japan Earthquake and nuclear disaster memorial museum*. Retrieved from www.fipo.or.jp/lore/en/training

Genpatsu jūmin tōhyō yorihiroku koe wo kike [Nuclear referendum: Citizens' voices must be heard]. (2020, June 16). *Tokyo Shimbun*. Retrieved from www.tokyo-np.co.jp/article/35785

Haas, R., Thomas, S., & Ajanovic, A. (2019). The historical development of the costs of nuclear power. In R. Haas, L. Mez, & A. Ajanovic (Eds.), *The technological and economic future of nuclear power* (pp. 97–116). Wiesbaden: Springer VS.

Haddad, M. A. (2012). *Building democracy in Japan*. New York: Cambridge University Press.

Harris, R. (2013, December 13). Environmentalists split over need for nuclear power. *NPR*. Retrieved from www.npr.org/2013/12/17/251781788/environmentalists-split-over-need-for-nuclear-power

Japan Center for Economic Research. (2019, July 3). *Accident cleanup costs rising to 35–80 trillion yen in 40 years*. Retrieved from www.jcer.or.jp/english/accident-cleanup-costs-rising-to-35-80-trillion-yen-in-40-years

Keeling, A., & Sandlos, J. (2009). Environmental justice goes underground? Historical notes from Canada's northern mining frontier. *Environmental Justice*, 2(3), 117–125. https://doi.org/10.1089/env.2009.0009

Kinefuchi, E. (2018). Production of the internal other in world risk society: Nuclear power, Fukushima, and the logic of colonization. In J. A. Drzewiecka, A. Jolanta, & T. K. Nakayama (Eds.), *Global dialectics in intercultural communication: Case studies* (pp. 265–286). New York: Peter Lang.

Kinsella, W. J. (2015). Rearticulating nuclear power: Energy activism and contested common sense. *Environmental Communication*, 9(3), 346–366. https://doi.org/10.1080/17524032.2014.978348

Kinsella, W. J. (2019). *Toxic temporalities: Prediction, purification, and perseverance of nuclear waste*. Paper presented at the Conference on Communication and Environment, Vancouver, Canada, June 17–21, 2019.

Kobayashi, Y. (2011). *Malfunctioning democracy in Japan: Quantitative analysis in a civil society*. Lanham, MD: Lexington Books.

Kokunai no genpatsu wo kongo dōsubekika? Genpatsujiko 10nen NHK yoron chōsa [What should be the future of Japan's nuclear power? NHK's public poll on the 10-year anniversary of the nuclear accident]. (2021). *NHK*. Retrieved from https://www3.nhk.or.jp/news/html/20210302/k10012893841000.html

Laclau, E. (2001). Democracy and the question of power. *Constellations*, 8(1), 3–14. https://doi.org/10.1111/1467-8675.00212

Laclau, E., & Mouffe, C. (1985/2001). *Hegemony and socialist strategy: Towards a radical democratic politics* (2nd ed.). London: Verso. (Original work published 1985).

Lovins, A. (2011, April 8). *Would the world be better off without nuclear power?* The Rocky Mountain Institute. Retrieved from https://rmi.org/insight/would-the-world-be-better-off-without-nuclear-power/

Makiuchi, S. (2021, March 13). Denshōkan, shimin no koe de henkaku unagase [Improve the memorial museum through citizens' voices]. *Seikei Tōhoku*. Retrieved from https://note.com/seikeitohoku/n/n974295eacda4

Malin, S. A. (2015). *The price of nuclear power: Uranium communities and environmental justice*. New Brunswick, NJ: Rutgers University Press.

Moore, M. T. (2018, September 5). Nuclear plant closures bring economic pain to cities and towns. *Pew*. Retrieved from www.pewtrusts.org/en/research-and-analysis/blogs/stateline/2018/09/05/nuclear-plant-closures-bring-economic-pain-to-cities-and-towns

Morales, L. (2016). For the Navajo nation, uranium mining's deadly legacy lingers. *NPR*. Retrieved from www.npr.org/sections/health-shots/2016/04/10/473547227/for-the-navajo-nation-uranium-minings-deadly-legacy-lingers

Moss, S. (2014, March 14). James Lovelock: 'Instead of robots taking over the world, what if we join with them?' *The Guardians*. Retrieved from www.theguardian.com/environment/2014/mar/30/james-lovelock-robots-taking-over-world

Natural Resource Defense Council. (2012). *Nuclear power's dirty beginning: Environmental damage and public health risks from uranium mining in the American West.* Retrieved from www.nrdc.org/sites/default/files/uranium-mining-report.pdf

Otmazgin, N., Ben-Rafael Galanti, S., & Levkowitz, A. (2014). Introduction: Japan's multilayered democracy. In S. Ben-Rafael Galanti, N. Otmazgin, & A. Levkowitz (Eds.), *Japan's multilayered democracy* (pp. 9–26). Lanham, MD: Lexington Books.

Plumer, B. (2017, July 31). U.S. nuclear comeback stalls as two reactors are abandoned. *The New York Times.* Retrieved from www.nytimes.com/2017/07/31/climate/nuclear-power-project-canceled-in-south-carolina.html

Plumer, B., Fountain, H., & Albeck-Ripka, L. (2018, April 18). Environmentalists and nuclear power? It's complicated. *The New York Times.* Retrieved from www.nytimes.com/2018/04/18/climate/climate-fwd-green-nuclear.html

Plumwood, V. (1993). *Feminism and the mastery of nature.* London: Routledge.

Plumwood, V. (2002). *Environmental culture: The ecological crisis of reason.* London. Routledge.

Plumwood, V. (2008). Shadow places and the politics of dwelling. *Australian Humanities Review, 44,* 139–150.

Rockström, J., Steffen, W., Noone, K., Persson, Å., Chapin, F. S., Lambin, E. F., . . . Foley, J. A. (2009). A safe operating space for humanity. *Nature, 461*(7263), 472–475. https://doi.org/10.1038/461472a

Sanchez, C. L. (1994). Animal, vegetable, and mineral: The sacred connection. In C. J. Adams (Ed.), *Ecofeminism and the sacred.* New York: Continuum.

Shiva, V. (2005). *Earth democracy: Justice, sustainability, and peace.* Berkeley, CA: North Atlantic Books.

Smith, D. M. (2018). *Dynasties and democracy: The inherited incumbency advantage in Japan.* Stanford, CA: Stanford University Press.

Steffen, W., Richardson, K., Rockström, J., Cornell, S. E., Fetzer, I., Bennett, E. M., . . . Sörlin, S. (2015). Planetary boundaries: Guiding human development on a changing planet. *Science, 347*(6223). https://doi.org/10.1126/science.1259855

Switzerland votes to phase out nuclear power. (2017, May 21). *BBC News.* Retrieved from www.bbc.com/news/world-europe-39994599

The International Atomic Energy Agency. (2016). *Nuclear power and sustainable development.* Retrieved from https://www-pub.iaea.org/MTCD/Publications/PDF/Pub1754web-26894285.pdf

Tierney, S., & Bird, L. (2020). *Setting the record straight about renewable energy.* Retrieved from www.wri.org/insights/setting-record-straight-about-renewable-energy

United States Nuclear Regulatory Commission. (n.d.). *Backgrounder on radioactive waste.* Retrieved from www.nrc.gov/reading-rm/doc-collections/fact-sheets/radwaste.html

Wade, W. (2019, June 14). Americans are paying more than ever to store deadly nuclear waste as plants shut down. *Bloomberg.* Retrieved from www.bloomberg.com/news/articles/2019-06-14/u-s-bill-to-store-nuke-waste-poised-to-balloon-to-35-5-billion

Watanabe, J. (2020, October 25). Genshiryoku saigai denshōkan de kangaeru: Matome [A summary of a discussion about the nuclear power disaster memorial museum]. *Café de Logos.* Retrieved from https://blog.goo.ne.jp/cafelogos2017/e/3092d2ec9300b6796d93862bd30ed685

World Nuclear Association. (2017). *Radioactive waste – Myths and realities*. Retrieved from www.world-nuclear.org/information-library/nuclear-fuel-cycle/nuclear-wastes/radioactive-wastes-myths-and-realities.aspx

World Nuclear Association. (2021, May). *Covid-19 Coronavirus and nuclear power*. Retrieved from www.world-nuclear.org/information-library/current-and-future-generation/covid-19-coronavirus-and-nuclear-energy.aspx

Yamashita, M. (2021, February 14). Kataribe ni hihan mitomezu: Genshiryoku saigai denshōkan rupo [Criticisms prohibited at the storytelling section: A report of the nuclear disaster memorial museum]. *Nishi Nihon Shimbun*. Retrieved from www.nishinippon.co.jp/item/n/692772/

Index

Note: Page numbers in *italics* indicate a figure on the corresponding page. Page numbers followed by "n" indicate a note.

Printed in the United States
by Baker & Taylor Publisher Services